AF342515

β-Casomorphins

Mohammad Raies Ul Haq

β-Casomorphins

A1 Milk, Milk Peptides and Human Health

 Springer

Mohammad Raies Ul Haq
Department of Biochemistry
Sri Pratap College, Cluster University Srinagar
Srinagar, Jammu and Kashmir, India

ISBN 978-981-15-3456-0   ISBN 978-981-15-3457-7   (eBook)
https://doi.org/10.1007/978-981-15-3457-7

This Springer imprint is published by the registered company Springer Nature Singapore Pte Ltd.
The registered company address is: 152 Beach Road, #21-01/04 Gateway East, Singapore 189721,
Singapore

Preface

This leading-edge composition entitled *β-Casomorphins* comprehensively describes the link between a type of milk we usually consume and its correlations with various health complications. During the early 1990s, a correlation between A1 milk intake and increased risk for incidence of disorders was set up on the basis of ecological correlations. This was further strengthened by animal trials, human case-control studies, and few in vitro findings. The difference between A1 and A2 milk is the presence of histidine and proline, respectively, at position 67 of β-casein. This mutation leads to the release of a seven amino acid peptide called β-casomorphin-7 (BCM-7) from the former that is actually a hypothetical risk factor. BCM-7 is an opioid peptide with morphine-like activity that binds mu (μ) receptors. A2 milk is high in the cattle breeds of Asia, Africa, and parts of Southern Europe. A1 milk is common in Europe, the United States, and New Zealand. The Indian indigenous cattle yield A2 milk and Indian crossbred cattle yield both milks, but still with higher A2 milk type.

In this book, I have tried to bring together scientific information from more than 100 research and review articles. This book describes cow milk and its composition, the status of A1/A2 milk hypothesis, and the structure, properties, and activities of BCMs. Moreover, correlations between consumption of A1 milk and type 1 diabetes mellitus, cardiovascular diseases, and neurological manifestation are also given comprehensively. A critical analysis by EFSA, American Nutritionists, and Truswell is also provided for the interest of readers. Furthermore, I have been working in the same research area since 2009 and established a correlation between consumption of A1 milk and BCM-7 with gut inflammation in mice models. We also screened animals (Indian indigenous and crossbred cattle) with PCR-ACRS. Additionally, we checked the release of these peptides from A1 and A2 β-casein with SGID.

I hope this composition will be useful for agricultural scientists in general and dairy sector in particular. This is quite interesting and amazing not only because of an establishment of correlation between nutrition and human health but due to the courageous criticism, value added corporate business, and consumer choices.

The final message of this book for readers and researchers is as follows:

The ecological, case-controls, in vivo, in vitro, biochemical, pharmacological, and immunological studies are strong enough to establish a correlation regarding A1 "like" milk consumption. The criticism cited by American Nutritionists, EFSA, and Truswell cannot be overlooked. The present status of the hypothesis is not in a position to make final consumer health recommendations regarding this issue. Therefore, more mechanistic studies are needed at cellular and molecular levels with explored signal cascade pathways. Overall, the hypothesis is potentially important and possibly significant. However, verified and authenticated research with reproducible results is needed to make final recommendations and guidelines regarding world's dairy sector and consumer health.

Srinagar, India   Mohammad Raies Ul Haq

Acknowledgments

Searching literature and subsequently researching in the field of A1 milk and then exploiting this experience in writing this book, largely in winter and summer vacations, at nights, and over a period of 1 year, have been an awesome but challenging experience.

In this endeavor, first and foremost I thank Almighty Allah who retained me in circumstances more favorable than acquiring this successful attempt. He (Allah) bestowed upon me with the lovely family (Dad, Mom, Wife, Sister, Dr. Arshid, Juwariya, Tahoora, and Faaiz) as a sounding board at both favorable and critical times.

I acknowledge Dr. Rajeev Kapila (Principal Scientist, NDRI-ICAR-Karnal, Haryana) as my teacher and mentor, who was particularly more helpful in teaching me some of the main biochemistry behind A1 milk story. He gifted me with his valuable time and research funding that started from preparation of synopsis to PhD thesis writing and research publications. This work would have been impossible in his absence.

I acknowledge Professor (Dr.) Bashir Ahmad Ganie (Director CORD, University of Kashmir) under whose mentorship I completed my Post-Doctoral fellowship (funded by SERB); the research was absolutely related to this book composition.

I acknowledge Mr. Khursheed, Dr. Shafi, Dr. Gulzar, Dr. Iqbal, Dr. Shabir, Dr. Showkeen, Dr. Sheraz, Dr. Danish, and Dr. Ahmad for the entertainment they provided that was subsequently essential for the sustenance of balanced and sound mind.

The composition of this book would have been impossible without some exceptional research work by many eminent scientists who made the discoveries on which this book is based. I corresponded with one or two such scientists, and the remaining are known to me through their outstanding research.

I specially thank Dr. Keith Woodford and Dr. Andrew Clark who have made special contributions in this field.

Moreover, the critics like Goldberg, Truswell, and DATEX Working Group (EFSA) are also highly acknowledged for their valuable commentaries and critical analysis.

Through the medium of this book I did not try to make any nutritional recommendation, but it is a collection of literature and my research expertise both in favor and against the A1 milk hypothesis. The supported data and the critical analysis are both highly acknowledged.

I tried to write this book with the hope that it is free of scientific errors. Nevertheless, it is just a hope; science is occasionally free of errors and omissions. The development of a scientific base is through repudiation of the principles that orthodox perception articulates are correct. The errors in the book are my responsibility.

I welcome suggestions and critical analysis for the next edition of this book.

About the Book

This present book entitled *β-Casomorphins* is a marvelous effort internationally to show a correlation between A1 cow milk consumption and increased risk for incidence of various diseases. It describes A1/A2 milk hypothesis and reveals how a single nucleotide mutation from proline (A2) to histidine (A1) leads to release of BCM-7 from only A1 variant and hence considered a hypothetical "risk" element. The purpose of this composition is to discuss the current status, evidences in favor, and the strong criticism against the hypothesis comprehensively. The purpose of this book is neither to favor (or discourage) any company or allied corporate sector nor to make any consumer health recommendation. It is just a collection of scientific literature and my opinions gained through research expertise in this field. To the best of my knowledge, I think the present status of the hypothesis is not in a position to make any public or consumer health recommendation. The need of the hour is to do deeper research at cellular and molecular levels and to explore signal cascade pathways in humans. The budding scientists working on food safety issues in general and dairy sector in particular are the best audience of this book. It will decipher great interest to professional and nonprofessional science graduates and PhD scholars working in animal and human sciences. The uniqueness of the book is in its structure that elaborately discusses the foundation, genetics, and current status of A1 milk hypothesis and its human health implications (biochemistry, pharmacology, case-controls, ecological, and human studies). This book incorporates updated literature, recent developments, and advances in this domain to date. For clear and easy comprehension, self-explanatory color figures, illustrations, flowchart diagrams, and tables have been incorporated.

Contents

About the Author

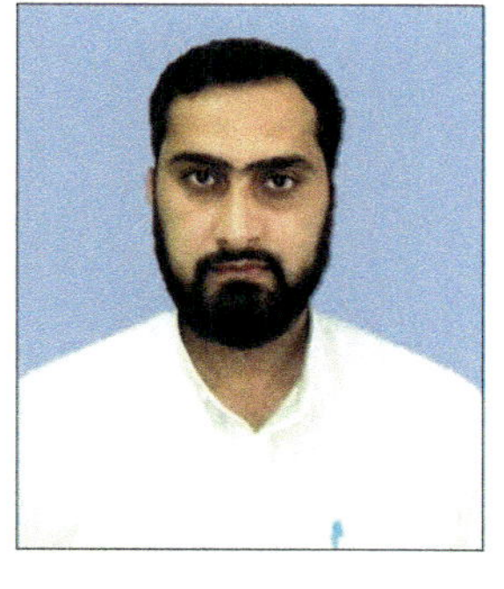

Mohammad Raies Ul Haq holds a PhD in Biochemistry from the National Dairy Research Institute (NDRI)-Indian Council of Agricultural Research (ICAR) and currently serves as a faculty member (Assistant Professor, Biochemistry) at Sri Pratap College, Cluster University Srinagar. He is an editorial board member of the *Journal of Dentistry and Oral Disorders* and of *Acta Scientific Nutritional Health*. A member of the Society of Biological Chemists and Indian Science Congress, Dr. Raies has several years of experience in the areas of teaching, research, and student counseling. He has worked as a lecturer in the Department of Biochemistry at the University of Kashmir. Dr. Raies has several research publications in reputed international journals.

Abbreviations

ACRS	Amplification-created restriction site
ALTE	Acute life threatening events
ASDs	Autistic spectrum disorders
AS-PCR	Allele-specific polymerase chain reaction
BB	Biobreeding
BCM	β-casomorphin
BCM-5	β-casomorphin-5
BCM-7	β-casomorphin-7
BCMIR	β-casomorphin immunoreactive
BSA	Bovine serum albumin
CD	Cluster of differentiation
CEP	Cell envelope proteinases
CHD	Coronary heart disease
CVD	Cardiovascular disease
DPP IV	Dipeptidyl peptidase IV
EFSA	European Food Safety Authority
ELISA	Enzyme-linked immunosorbent assay
EMC	Enzyme-modified cheddar cheese
FAO	Food and Agriculture Organization
FFDCA	The U.S. Federal Food, Drug, and Cosmetic Act
GABA	Gamma-aminobutyric acid
GER	Gastric emptying rate
GIP	Gastrin inhibitory peptide
GITT	Gastrointestinal transit time
GLUT-2	Glucose transporter 2
GPI	Guinea-pig ileum
HDL	High-density lipoprotein
HLA	Human leukocyte antigen
ICAR	Indian Council of Agricultural Research
IEC	Isoelectric casein
IF	Infant formula
IHD	Ischemic heart disease

ISI	Ideasphere incorporated
IUPHAR	International Union of Basic and Clinical Pharmacology
KF	Karan Fries
KS	Karan Swiss
LAB	Lactic acid bacteria
LAP	Leucine aminopeptidase
LDL	Low-density lipoprotein
LPL	Lamina propria lymphocytes
MALDI-TOF	Matrix-assisted laser desorption/ionization-time of flight
MCP-1	Monocyte chemotactic protein-1
MMR	Measles-Mumps-Rubella
MPO	Myeloperoxidase
MS	Mass spectroscopy
MUC2	Mucin2
MVD	Mouse vas deferens
NDRI	National Dairy Research Institute
NOD	Non-obese diabetic mice
oxLDL	Oxidized low-density lipoprotein
PBMCs	Peripheral blood mononuclear cells
PBS	Phosphate buffer saline
PO	Prolyl oligopeptidase
RFLP	Restriction fragment length polymorphism
ROS	Reactive oxygen species
RP-HPLC	Reverse phase-high pressure liquid chromatography
RR	Relative risk
RT	Retention time
SCFA	Short-chain fatty acids
SCIT	Subtle cognitive impairment test
SDS-PAGE	Sodium dodecyl sulfate-polyacrylamide gel electrophoresis
SGID	Simulated gastrointestinal digestion
SIDS	Sudden infant death syndrome
SSC	Somatic cell count
SSCP	Single-stranded conformation polymorphism
T1D	Type 1 diabetes mellitus
TLR	Toll like receptor
UHT	Ultra heat treatment
USDA	United States Department of Agriculture
VLDL	Very low-density lipoprotein
WHO	World Health Organization

Chapter 1
Cow Milk

Abstract Cow milk is a colloidal solution that contains numerous biomolecules essential for growth and maintenance. Caseins in cow milk are phosphoproteins divided into α, β, κ-forms. β-casein has 12 genetic variants that are represented by A1, A2, A3, B, C, D, E, F, H1, H2, I, G letters. The mutation on exon VII on sixth chromosome of cow β-casein gene leads to replacement to histidine from proline at position 67. It leads to the division of milk into A1 and A2 "like". A1 "like" milk (histidine, 67) includes B, C, F and G alleles while A2 "like" milk (proline, 67) further includes A3, D, E, H1, H2 and I alleles. Some of the Western cattle breeds (Angus, Ayrshire, Black Pied and Braham) demonstrate higher frequency of A1 and some depict lower A1 (Brown Italian, Brown Swiss and Guernsey) allele frequency of β-casein gene in cow milk. The Indian indigenous cows show higher A2 allele frequency while the crossbred cows demonstrate both alleles, still with higher A2 allele frequency.

1.1 Introduction to Cow Milk

Milk is a source of essential biomolecules including carbohydrates (lactose) as an immediate source of energy, premium quality proteins (complete proteins) for growth and maintenance especially for young ones, and lipids (storage and energy purposes). It makes a considerable contribution to the much needed nutrient consumptions for minerals like calcium, magnesium, sodium, iron, potassium and vitamins including riboflavin, vitamin B12 and pantothenic acid. Milk and other dairy products are nutrient-rich diets and their intake supplements the plant-derived nutritional components. The composition, color and flavor of cow milk changes with the genetic make-up (breed), age of dairy animals, diet formulations, stages of lactation, farming system, environmental alterations (hot and cold climates) and seasons (FAO 2019a, b).

Cow milk is a colloidal solution or emulsion composed of fat droplets suspended in universal solvent, water. The other essential elements include carbohydrates like lactose, glucose, galactose and some other oligosaccharides. Lactose is the most abundant sugar (5%) that is in the dissolved form in the milk. The other nutritional

© Springer Nature Singapore Pte Ltd. 2020

M. R. Ul Haq, *β-Casomorphins*, https://doi.org/10.1007/978-981-15-3457-7_1

elements include proteins aggregates (3.3%), the inorganic compounds in the form of minerals (0.7%), salts and the vitamins essential to the body. The cow milk contains about 3.6% fat (Jost 2002).

Cow milk is a source of nutrition for infants and adults and the daily intake is country dependent. The pH of the cow milk varies from 6.4 to 6.8 that change with time during storage. The nutrient composition of milk is shown in Table 1.1 as per United States Department of Agriculture (USDA), Agricultural Research Service, USDA Branded Food Products Database.

The fat in the milk exists in the form of globules that are surrounded by membranes. The fat globules mostly (97–98%) consists of a glycerol molecules esterified with three fatty acids as triacylglycerols, however, some other lipids like monoacylglycerols, diacylglycerols, cholesterol either in free or esterified form,

Table 1.1 Nutrient data for cow milk

United States Department of Agriculture (USDA), Agricultural Research Service, USDA Branded Food Products Database, 2018		
Nutrients	Units	Values/100 mL milk
Proximates		
Energy	kcal	108.0
Protein	g	3.330
Total lipid (fat)	g	4.580
Total carbohydrates	g	13.75
Lactose	g	5.260
Fiber	g	0.400
Minerals		
Calcium, Ca	mg	125.0
Iron, Fe	mg	0.300
Potassium, K	mg	175.0
Sodium, Na	mg	112.0
Magnesium, Mg	mg	10.00
Vitamins		
Vitamin C, total ascorbic acid	mg	0.000
Vitamin A	IU	208.0
Vitamin B1	mg	0.044
Vitamin B2	mg	0.183
Vitamin B12	µg	0.45
Choline	mg	14.30
Vitamin D	IU	2.000
Lipids		
Fatty acids, total saturated	g	2.920
Fatty acids, total trans	g	0.000
Cholesterol	mg	15.00
Amino acids	g	0.017–0.648

Units: *kcal* kilocalories, *g* grams; *µg* micrograms; *IU* international units free fatty acids and phospholipids also do exist. The membranes of these globules are made up of phospholipids and proteins. The advantage of the membrane is the maintenance of the globules in their individual form preventing them from coalescing. Moreover, these membranes prevent the globules from degradation by enzymes present in milk. The composition of fat in the cow milk varies and depends on the genetic composition of breeds, nutritional status and locational stages (Fox 1995).

Cow milk consists of about 3.3% proteins. The latter are divided into two broad categories i.e., caseins and whey proteins that differ in their amino acid composition and physical characteristics. The former exists as casein micelle while as the whey proteins remain in the whey (serum) of the cow milk. These consist of approximately α- and β-lactglobulins 20% and 50% respectively, blood serum albumin, antibodies, iron containing lactoferrin and transferrin, and some enzymes. The casein proteins are phosphoproteins, therefore, precipitate at pH 4.6, while as the whey proteins are deficient in phosphorus that is the reason these protein remain in solution in milk at this pH. The amount of proteins found in 1 L of cow milk is between 30–35 g. The casein protein is further divided into α, β, κ-caseins. It provides all the essential amino acids to the body, therefore regarded as complete protein (Farkye 2003).

The cow milk consists of inorganic elements in the form of minerals (0.7%) in their cationic or anionic form. These ions may exist either in their free form or as combined mineral salt. The various minerals in milk include calcium, sodium, chloride, phosphate, magnesium, potassium and citrate. The concentration of these elements vary in cow milk, however, the strength may vary between 5–40 mM. The mineral salts especially calcium phosphate has a greater tendency to interact with casein proteins of cow milk. The casein micelles in cow milk comprises of 67% of the calcium, 44% of the phosphate and 35% of the magnesium. The association of calcium and phosphate with the casein micelle doesn't disturb their nutritional availability. Milk contains small amounts of copper, iron, manganese, and sodium and is not considered a major source of these minerals in the diet (Flynn and Cashman 1997).

Cow milk also contains other organic compound known as vitamins that are essential to the human body as energy releasing factors or hematopoietic elements. Cow milk contains thiamine (B1), riboflavin (B2), niacin (B3), pantothenic acid (B5), vitamin B6 (pyridoxine), vitamin B12 (cobalamin), vitamin C, and folate. However, it is considered as a good source of vitamin B1, B2 and B12. It comprises of fat soluble vitamins A, E, and K, nevertheless, milk is not considered as a good source of vitamin D (Öste et al. 1997).

The other miscellaneous substances present in the cow milk in its raw form include white blood cells, mammary gland cells and some bacteria. Cow milk is also a source of numerous enzymes that are present in its native form or added during airborne bacterial contamination, through bacterial fermentation, or in somatic cells present in milk. These enzymes include lipoprotein lipase, proteases (plasmin), alkaline phosphatase (marker of pasteurization), and lactoperoxidase (heat-stable) (Farkye 2003).

1.1.1 Milk Production

Globally, approximately 150 million households are involved in the production of milk. In developing countries, the milk production is done by smallholders and they use it as an important source of cash income, nutrition or food security. For small-scale milk producers, this delivers comparatively rapid earnings. During the last few decades, the contribution of the developing countries for milk productions has increased relatively. It is reported that the increase is due to the number of animals rather than rise in productivity per head. However, for some developing countries, milk production is still rate limiting due to the poor feed quality, inefficient medical facilities and inferior genetic make-up of dairy animals. The other predominant factor in the developing compared to developed countries for lower milk production is the environmental factor, i.e., hot and humid climates that aren't favorable for milk production. Among developing countries, few have a long history of dairy production, the milk and other dairy products have therefore a significant role in human health and nutrition. Other developing countries have established the dairy farming and production more recently. The countries with long history of milk production are located in the Indian subcontinent, situated in the Mediterranean and Near East regions, the savannah provinces of West Africa, the highlands of East Africa and parts of South and Central America. On the other hand, the countries that have recently contributed in sector include Asia (Southeast) and the tropical regions that have high ambient temperatures and humidity. The following are important considerations regarding milk production around the world (FAO 2019a, b).

- Globally, during the last three decades, milk production has improved by more than 58% (522–828 million from 1987 to 2017).
- India occupies the first position in milk production with a share of 21% in the global production followed by USA, China, Pakistan and Brazil.
- From 1970, the milk production has increased drastically in South Asia that is the chief driver of milk production growth among the developing countries.
- In African countries, the rate of milk production is developing more slowly; this is due to poverty and adverse climatic conditions.
- The highest surpluses of milk are found in New Zealand, the USA, Germany, France, Australia and Ireland.
- The countries that have highest milk deficits include China, Italy, the Russian Federation, Mexico, Algeria and Indonesia.

1.1.2 Milk Processing

Quite evidently, the shelf-life of milk is short and hence needs careful processing and storage. Milk is susceptible to spoilage because it provides an excellent medium and in turn offers great opportunities for the growth of bacteria (pathogen particularly). The growth of these pathogens leads to a negative impact on consumers in terms of many illnesses. The proper milk processing leads to preservation of milk from days to months through reduction of milk-borne ailments. There are many conventional and advanced techniques available that are employed for such purposes. Cooling is one of the methods used to increase shelf-life of raw milk. The other methods include fermentation or pasteurization. The latter involves heat treatment of milk to the extent that pathogenic bacteria can't grow and hence prevents humans from incidence of any health complications. Milk can be manufactured to other dairy products including butter, cheese and ghee that have high nutritive value in addition to enhanced shelf-lives. These processed dairy products enable the producers to earn high income compared to selling of raw milk. Further, it creates great opportunities for the raw milk to reach the urban and far-flung areas. This ultimately helps whole society to generate off-farm jobs from milk collection through transportation and processing to marketing (FAO 2019a, b).

1.1.3 Milk Products

The "milk product" according to the Codex Alimentarius is defined as a "product of milk that is obtained after proper processing, it may possess some additives or preservatives essential for the processing". The dairy products obtained from milk after proper processing is country and region dependent, depends on dietary habits, the processing methods available in the region, the market demand and societal and cultural conditions. The per capita consumption of milk and other dairy products is higher in developed compared to developing countries. Nevertheless, this mandate rises day by day in the latter with growing incomes and population, urbanization and changes in diet habits. In East and Southeast Asia, mostly in population-dense regions like China, Indonesia and Viet Nam, this trend is pronounced enormously. The production of dairy products from whole/raw milk enables the producers and other peri-urban people to increase livelihoods. The milk in its liquid form by volume is reported as the most consumed milk product all over the developing world. Conventionally, there is a great demand for liquid milk in urban area compared to rural areas which prefer fermented milk products. Following are the interesting considerations regarding the milk products.

- Worldwide, approximately more than six billion people take milk and other dairy products.
- The majority of these consumers inhabit in developing countries.
- Starting from 1960s, although the consumption of milk per capita has increased (twofold), nevertheless, it still lags behind meat (tripled) and egg consumption (fivefold).
- During the last two decades, milk intake per capita has decreased in sub-Saharan Africa.

- The milk consumption per capita is high in Argentina, Australia, Europe, Israel, North America and Pakistan; medium in India, Iran, Japan, New Zealand, and low in Viet Nam, Senegal, Africa and East and Southeast Asia
- Approximately, about 50% of milk in India is consumed on-farm.

1.2   Cow Milk Proteins

Cow milk consists of 3.3% total protein. It possesses all the nine essential amino acids that are required for growth and maintenance in humans, therefore regarded as complete proteins. It is reported that about 60% of the amino acids that are needed for protein synthesis in humans are obtained from cow milk. However, the total protein content in the cow milk depends on its genetic composition that differs from breed to breed. The milk proteins are divided into two broad classes i.e., caseins and whey proteins.

Milk proteins may be degraded with the use of enzymes or when exposed to sun light. However, the principle component of milk protein degradation is through enzymes called proteases. These enzymes are either present in the native milk, added through airborne bacterial contamination, intentionally added for fermentation or present in somatic cells of milk. The application of proteases in milk is referred to as desirable when used for preparation of dairy products like yogurt or cheese. Plasmin is the most commonly used desirable enzyme for cheese production that produces desirable flavors and texture in cheese. However, the undesirable proteolysis results in milk with off-flavors and poor quality. Methionine and cystine present in milk are sensitive to light and therefore are susceptible to degradation with exposure to light. The result is the production of off-flavor in the milk and further loss of nutritional quality for these two amino acids. The whey is also called "serum" of the milk and the proteins that remain dissolved in it when casein is precipitated are referred as whey proteins. These include β-lactglobulin (50%), α-lactalbumin (20%), albumin, antibodies, iron containing lactoferrin and transferrin, and other protein in small amounts including some enzymes. The whey proteins differ in their amino acid composition, genetic variations and ultimately in their structure and biological properties and functions. Unlike caseins, these proteins contain huge amounts of sulphur compared to phosphorus in caseins. The presence of sulphur in these proteins actually resides in the sulphur containing amino acids cysteine and methionine. These amino acids residues result in the formation of disulphide bridges causing the chains to form compact spherical shape. However, the denaturing conditions lead to the loss of this compact structure. Denaturation of whey proteins improves the texture of the yogurt as the amount of the water molecules that bind these proteins increases. This property is also exploited for the formation of other specialized whey protein constituents with exceptional well-designed properties for use in foods. For example, addition of whey proteins to specialized sausage and meat products to increase absorption of water molecules and in turn improvement in the texture of these products. The whey proteins have multifunction including carrier of vitamin A (β-lactglobulin), lactose synthesis (α-lactalbumin), animals defense system (immunoglobins), absorption and transport of iron (lactoferrin and transferrin).

1.3 Cow Milk Caseins

The caseins are phosphoproteins and have isoelectric pH of 4.6 and therefore precipitate at this pH. On the other hand, the whey proteins aren't phosphoproteins and hence don't precipitate at this pH and exist as soluble protein in whey. The principle of curd formation, coagulation or cheese production is based on the mechanism of casein precipitation at lower pH (4.6). Approximately, 82% of cow milk proteins are caseins and remaining 18% comprises of whey proteins. Cow milk comprises of four casein proteins referred as αs1 (39–46% of total caseins), αs2 (8–11%), β (25–35%), and κ (8–15%). The γ-casein also does exist, however, considered a degradation product of β-casein. The individual casein proteins differ in amino acid composition, genetic variations and hence functional attributes (Roginski 2003). The caseins exist in milk in the form of micelles suspended in the water phase of milk. These micelles are composed of different negatively charged caseins and held together by the calcium phosphate bridges on the inside. These micelles are spherical in shape and their size varies from 0.04 to 0.3 μm in diameter. The structure of the micelle is porous in nature and therefore allows free passage of water in and out of this structure. These micelles are dynamic and stable structures and don't settle out of solution. The high content of amino acid proline in casein confers it a random and open structure. The presence of negatively charged phosphate allows the casein to bind with the calcium to form calcium phosphate salts. The high phosphate content in milk enables it to have much more calcium compared to when calcium was dissolved in solution.

1.4 β-Casein Variants

The cow milk β-casein protein is composed of 209 amino acids and this exits in a single folded chain. For the first time in 1961, the polymorphism of milk protein β-casein was observed in Jersey and Guernsey cows by Aschaffenburg (Aschaffenburg 1961). It was reported that β-casein exists in three genetic variants represented by A, B and C. He determined these polymorphs by paper electrophoresis using urea as the denaturant at a concentration of 6 M in a citrate-phosphate buffer at a pH 7.5. It was followed by the observations of Peterson and Kopfler in 1966 who analysed the sub-variants of A β-casein as A1, A2 and A3, these workers used polyacrylamide gel electrophoresis in acid urea for this analysis. These three genetic variants were found to differ in the primary amino acid structure in having basic amino acid residues (positively charged amino acids) (Peterson and Kopfler 1966).

A1 variant was the first genetic variant that was determined, A2 was revealed second, so on and so forth. Literature has clearly confirmed that the most common forms of A β-casein genetic variants are A1, and A2. It is well-established that A2 β-casein genetic variant was the original variant found during evolutionary periods. The advanced molecular and biochemical techniques including chromatographic analysis, isoelectric focusing and nucleic acid based PCR-RFLP, PCR-SSCP have revealed at least 12 genetic variants of β-casein (Table 1.2). These genetic variants are represented as A1, A2, A3, B, C, D, E, F, H1, H2, I, G and the order of alphabets and digits represent the sequence in which they were identified. The mutation on exon VII on sixth chromosome of cow β-casein gene leads to the nitrogenous base substitution from cytosine to adenine that in turn leads to amino acid replacement to histidine (codon; CAT) from proline (codon; CCT) at position 67 (Groves, 1969). The aforementioned mutation is of great importance as it leads to the division of milk into A1 "like" and A2 "like". The former type of milk includes A1 β-casein or it's like variants including B, C, F and G variants all with histidine at position 67 (−Tyr60-Pro61-Phe62-Pro63-Gly64-Pro65-Ile66-His67), however, there may be changes at other positions. A2 "like" milk further includes variants A3, D, E, H1, H2 and I all having proline at position 67, with changes at other positions in amino acids (−Tyr60-Pro61-Phe62-Pro63-Gly64-Pro65-Ile66-Pro67-) as shown in Fig. 1.1 (Raies et al. 2012a, b).

Table 1.2 Cow milk β-casein variants with genetic variations

β-casein variants	18	25	35	36	37	67	72	88	93	106	117	122	137	138
A2	S-p	R	S-p	E	E	P	E	L	Q	H	Q	S	L	P
A1						H								
A3						P				Q				
B						H						R		
C			S		K	H						R		
D	K					P								
E				K		P								
F						H								L
G						H					L			
H1			C			P		I						
H2						P	E		L					
I						P			L					

- A1//A2 "like" milk classification is based on mutation at 67th amino acid position
- Proline (P) (green) containing variants are A2 "like" and Histidine (H) (yellow) possessing variants represent A1 "like" milk
- The digits indicate the position of amino acids in the β-casein protein of cow milk
- The letters refer to one letter code for amino acids

1.4.1 Worldwide Distribution of A1/A2 β-Casein Allele Frequency in Cows

During the last six decades, there has been a great work in the field of the analysis of β-casein genetic polymorphism. However, the same research took a tremendous enthusiasm around 1990 when A1/A2 milk hypothesis was laid down. For the first time, in 1961, Aschaffenburg studied the genetic polymorphism of β-casein gene in Guernsey and Jersey cows. For such analysis he performed paper electrophoresis using urea (6 M) as the denaturating agent in presence of buffer (citrate-phosphate, pH 7.5). He found that β-casein gene in these cows has three variants and designated them as A1, B and C in order of decreasing mobility (electrophoretic). The work was later extended by Peterson and Kopfler in 1966 in Jersey, Brown Swiss and Ayrshire, these workers used polyacrylamide gel electrophoresis (PAGE) in presence of acid urea. It was surprising to note that A variant of β-casein further resolved into A1, A2, and B β-casein genetic variants. These individual A variants were found to differ in number of basic amino acid residues (positively charged amino acid, histidine) in their primary sequence. The three most significant species of the genus *Bos*, i.e. *Bos taurus* (taurines), *Bos indicus* (zebu) and *Bos grunniens* (yak) have been studied for the genetic polymorphism of β-casein gene. Nevertheless, comprehensive data is also available for taurines, this is due to the fact that these are the chief milk generating cows in the Western world. In the latter countries, studies have revealed the higher frequency of three genetic variants of β-casein i.e., A1, A2 and B. The data generated shows that there are many differences in β-casein gene polymorphism when it is related to numerous countries. It is believed that these variations may be due to local breeding policies either cross breeding or other targeted breeding for improvement of milk production traits. The literature depicted some intriguing trends like Normande, Jersey and Simmental breeds represented regularity in possessing a comparatively lesser A1 β-casein allele frequency. On the other hand these breeds, in comparison with the other breeds listed in this Table 1.3 (Normande, Jersey and Hereford) were found to have higher B β-casein allele frequency. Contrariwise, the Guernsey cows are reported to have almost entirely of A2

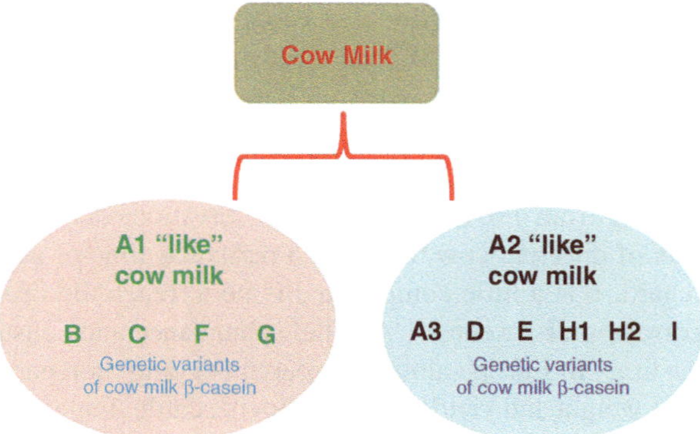

Fig. 1.1 Classification of cow β-casein variants into A1 and A2 "like" milk

Table 1.3 Genotype frequency A1/A2f β-casein in cattle breeds *c Rep*

Frequency of A1/A2 β-casein in cattle breeds (globally)				
Breeds	β-Casein allele frequency			References
	A1	A2	Other	
Guernsey	0.010	0.980	0.020	Aschaffenburg (1963)
Ayrshire	0.600	0.400	0.000	Aschaffenburg (1968)
Angus	0.950	0.000	0.050	Caldwell et al. (1971)
Jersey	0.070	0.570	0.360	McLean et al. (1984)
Holstein-Frisian	0.363	0.632	0.001	Lin and McAllister (1986)
Red Danish	0.710	0.230	0.060	Bech and Kristiansen (1990)
Simmental	0.231	0.673	0.095	Baranyi et al. (1993)
Brown Swiss	0.320	0.520	0.160	Ng-Kwai-Hang and Kim (1994)
Holstein	0.580	0.400	0.020	Aleandri et al. (1997)
Black pied	0.970	0.000	0.030	Lien et al. (1999)
Friesian	0.380	0.550	0.070	Boettcher et al. (2004)
Holstein-Frisian	0.402	0.598	0.000	Kaminski et al. (2007)
Swedish lowland cattle	0.340	0.600	0.060	Hallén et al. (2008)
Frequency of A1/A2 β-Casein in Indian Cattle Breeds Raies et al. (2013)				
Sahiwal	0.160	0.680	0.160	Dhanammal (2006)
Sahiwal, Tharparkar	0.013	0.987	0.000	Mishra et al. (2009)
Sahiwal, Tharparkar	0.000	1.000	0.000	Raies et al. (2012a, b)
Karan fries	0.208	0.792	0.000	
Karan Swiss	0.107	0.893	0.000	
Ongole	0.060	0.940	0.000	Ganguly et al. (2013)
Frieswal heifers	0.320	0.680	0.000	
Malnad Gidda	0.014	0.986	0.000	Ramesha et al. (2016)
Rathi, Sahiwal	0.000	1.000	0.000	Saran (2017)
Badri cattle	0.120	0.880	0.000	Dar et al. (2018)
Vrindavani	0.35	0.650	0.000	Kumar et al. (2019)
Sahiwal	0.060	0.940	0.000	

allele and A2A2 genotype. Gray, Hungarian Spotted, Brown Swiss and Brown Italian cows yield both the types of milk (A1 and A2 "like"). However, the occurrence of A2A2 genotype was found to higher frequency in comparison with A1A1 genotype. In case of Holstein and Friesian cows, the frequency of A1 or A2 allele that is dependent on the regions they inhabit, certain regions favour A1 allele, others A2 allele and several cows denote equal A1/A2 allele frequency. The data represented in the frequency distribution table (Table 1.3) is approximately of 60 years starting from 1961 to 2019. The observations depict an overall frequency distribution of different cow breeds at a particular time of genotype determination. In this regard, it is a little complicated issue to reach the final conclusion and establish a correlation between A1/A2 allele abundance and consumer population at a particular time. In Europe already research is going on regarding the improvement in carcass weight and yield of milk in bovine cattle. For example, reports have established that in Nordic countries the milk production has enhanced remarkably between

1985 and 1997, however, congruently there has been substantial alterations in the composition of breeds at the national level. Literature has clearly depicted that the production of milk in Finland before 1960 was mostly taken from Finn cattle (Western, Northern and Eastern). However, the scenario has completely changed in the current times where this yield is now available from Finnish Holstein-Friesian and Finnish Ayrshire. The same case has been found in European Countries where the composition of the milk producing dairy cattle has changed throughout the same time course. In this regard, looking at in the distribution of A1 and A2 β-casein alleles in different breeds of cattle, it may be concluded that change in cattle breed composition altered the frequency distribution of A1/A2 β-casein alleles of this gene. Contrariwise, this assumption can't be applied to the region/countries that have not allowed foreign intrusion of the cattle in their countries. For example, the Iceland hasn't permitted the importation of foreign cattle in their country; therefore the breed composition of the dairy cattle hasn't changed remarkably with A1/A2 β-casein allele frequency. It is quite possible there may a little variations in the gene pool of such breeds but that is due the selective practices to improve or maintain the gene pool for certain useful traits. However, to the best of our knowledge and literature available, it is observed that there aren't any variations in the A1/A2 β-casein variants composition in the cattle residing in Iceland. More recently in 2019 Cieślińska et al. (2019) screened the Polish Red Bulls and cows and observed the A2 frequency of 0.58 and 0.37, respectively (Cieślińska et al. 2019).

1.4.2 Distribution of A1/A2 β-Casein Alleles in Indian Cows

To the best of the literature we searched, it was found that Indian indigenous cattle possess predominantly A2 allele and A2A2 genotypes of milk protein β-casein. However, the crossbred cattle contain both A1 and A2 alleles with A1A1, A2A2 and A1A2 genotypes for β-casein milk protein. It is therefore evident that Indian cattle possess favoured variant of β-casein and ultimately a premium type of cattle milk. At the premier Dairy Institute of India i.e., National Dairy Research Institute (NDRI), Karnal-Haryana, Dhanammal in 2006 screened Sahiwal cattle for A1 and A2 alleles of β-casein. It was found that the cattle possessed higher A2 allele frequency (0.68) compared to A1 (0.16) and A1 "like" allele, B (0.16). This was followed by the frequency determination by Mishra and coworker at the other nearby reputed research Institute in the same city (National Bureau of Animal Genetic Resources) in 2009. The frequency of these alleles was checked in Sahiwal, Tharparkar, and buffalo (Murrah). The results indicated a predominance of A2 allele (0.987) in Indian Zebu cattle breeds while the river buffalo indicated only A2 allele. We also screened crossbred (Karan Fries and Karan Swiss) and indigenous cattle (Sahiwal and Tharparkar) in our laboratory at National Dairy Research Institute (NDRI) in 2012. We also found that indigenous cattle possessed entirely of A2 allele (1.0) and A2A2 genotype (1.0). Additionally, we couldn't find any homozygous A1A1 or heterozygous A1A2 condition in these cattle. However, in case of crossbred cattle, both alleles were determined, still A2 allele being in predominance. In Karan Swiss A1 and A2 allele frequency was observed to be 0.107 and 0.893, respectively and genotypic frequency of A2A2–0.786, A1A2–0.214 and A1A1–0.0. In Karan Fries, we found an allelic frequency of A2–0.792 and A1–0.208 with genotypic frequency of A2A2–0.709, A1A2–0.166 and A1A1–0.125. These crossbred cattle were developed in the same premier Institute with the exotic breeds including Holstein Friesian and Brown Swiss. The later breeds may therefore act as the carriers of the A1 allele as the indigenous cattle breeds of India (Sahiwal and Tharparkar) were found to lack A1 variant. In Indian breeds of cattle Ongole (Zebu cattle) and crossbred Frieswal (HF × Sahiwal), Ganguly and coworkers in 2013 found a remarkably very low frequency of A1 compared to A2 allele of β-casein of cow milk protein. In the southern campus of NDRI, Ramesha and coworkers in 2016 found that the frequency of A1 allele of β-casein of milk protein was very low in Malnad Gidda (0.014), Kasargod variety (0.042) and Jersey (0.077). However, an improved frequency of this allele (A1) was found in crossbred cattle Holstein Fresian (0.169) and Holstein Fresian crossbred males (0.294) respectively. In a similar attempt in 2017, Saran screened the indigenous cattle, Rathi, Tharparkar, Kankrej, Sahiwal and crossbred of Rathi (Rathi × Holstein Fresian). He observed 100% A2A2 genotype in Rathi, Tharparkar and Kankrej cattle. However, Sahiwal cattle showed heterozygous A1A2 and homozygous A2A2 genotypes. Nevertheless, the A2 allele frequency (0.933) dominated the A1 allele (0.06) frequency. More recently in 2018, the A1/A2 allele frequency was assessed in Badri cattle by using molecular biology technique of PCR-RFLP. The researchers established a high A2 allele (0.88) frequency with genotypic frequency for A2A2 (0.76) and A1A2 (0.24). The authors finally concluded that this breed of cattle is a potential source of premium A2 milk with preferential or favorable A2 variant of casein.

1.4.3   Methods of Screening β-Casein (A1/A2) Variants

The aforementioned discussion clearly establishes that there are at least 12 genetic variants of β-casein in various dairy cows. However, the concept of the present book according to the A1/A2 milk hypothesis focuses on A1 and A2 alleles of β-casein of the cow milk. From the last six decades, the researchers have tried their best for the evaluation of these alleles by different biochemical or molecular biology testing techniques. The techniques employed for this purpose are discussed below.

1.4.3.1   Paper Electrophoresis

It is used for the separation of charged molecules like amino acids or peptides and proteins. In this method, a small strip of filter paper is first moistened with the buffer and the ends of the strip are then immersed into two reservoirs at the opposite ends. The reservoirs filled with the buffer contain two electrodes. The samples with analytes are applied at the centre of the strip. This is followed by the application of high voltage and the charged molecules move according to their charge. The detection of the separated components is then done by different staining techniques depending on their chemical nature.

Aschaffenburg in 1961 used this technique for the separation of β-casein variants of dairy cows. He used urea as a denaturant and performed electrophoresis in citrate-phosphate buffer at a pH of 7.5. The three variants of β-casein A, B and C differed in the electrophoretic mobility. It was found that the A variant was the most common and found in all the studied breeds. Further, Ayrshires and Shorthorns were found to have no other variant. The variant B was observed in Jersey breeds and also in Guernseys and Friesians but with a low frequency and variant C was also observed with a low frequency in Guernsey.

1.4.3.2 Acid-Urea Polyacrylamide Gel Electrophoresis

Quite evidently, SDS and urea are the chemicals that are used in protein and nucleic acid denaturation. SDS-PAGE is significantly used in the analysis of proteins. However Urea-PAGE is used both for protein and nucleic acid analysis. SDS is a strong anionic detergent and besides denaturation gets hydrophobically added to proteins, one SDS to two amino acids on an average. In this regard charge: mass ratio remains the same, then the analytes move according to their molecular mass due to sieving effect of polyacrylamide gel. The urea-PAGE is usually performed at high concentration of urea typically 6 M. In this method, no addition of external charge is added, therefore charge: mass ratio is estimated by both the size and intrinsic charge on the analytes. Therefore, if two peptides or proteins exist in a single band in SDS-PAGE, they may be separated into different bands in Urea-PAGE. It is advantageous because it is easy to perform compared to isoelectric focusing.

The two scientists (Peterson and Kopfler) in 1966 used this technique for analysis of β-casein variants of cow milk. Unlike Aschaffenburg in 1961, they surprisingly observed that A1 allele could be further resolved into A1, A2 and B β-casein variants of cow milk. It was further elucidated that this deeper analysis was due to the different number of histidine residues (Basic or positively charged amino acids) in the primary structure of A1 β-casein variants (6 residues), A2 (5 residues) and A3 (4 residues).

1.4.3.3 PCR-Single Stranded Conformation Polymorphism (PCR-SSCP)

This is a molecular biology technique used for genotyping of organisms. This works on the principle that a single mutation in DNA strand leads to its conformational changes that in turn changes electrophoretic mobility in non-denaturing or partially denaturing conditions.

Whenever there is a single change in a nucleotide or base pair in a double-stranded DNA, it is very difficult to distinguish these DNA fragments with electrophoresis as they demonstrate similar physical properties on which these can be resolved. However, on denaturation, they behave differently, and adopt disparate conformations. These altered conformations move differently during agarose electrophoresis and is exploited for their separate in genotyping.

Numerous studies are available that exploited this technique for screening of dairy animals (Brown Swiss, Sahiwal, Czech Jersey, Buffalo) for A1/A2 alleles of β-casein of milk proteins.

1.4.3.4 PCR-Restriction Fragment Length Polymorphism (PCR-RFLP)

Another molecular biology technique exploited for screening of organisms for the alleles of a gene and determination of genotypes. This technique differentiates the two alleles only when there is a naturally occurring restriction digestion site in one allele of a gene but not in its alternative form. The fragment containing the restriction site is first amplified with the PCR-thermocycler. The product obtained from the PCR from the alternative forms of gene is incubated with the particular restriction enzymes having restriction site for one allele of a gene. On agarose gel electrophoresis, three patterns are obtained, the homozygous for one allele without restriction digestion yields a single band, homozygous for another allele with a restriction digestion yields two bands (large and a small DNA fragments) and the heterozygous form yields three bands. Many scientists employed this technique for determination of A1 and A2 allele frequency of β-casein gene in cattle and buffalo (Sahiwal, Tharparkar, and Murrah). More recently, in 2019 the investigators used the same technique for genotyping of Tharparkar and Frieswal cattle at Indian Veterinary Research Institute, a premier institute of Indian Council of Agricultural Research (ICAR).

1.4.3.5 PCR-Amplification Created Restriction Digestion Polymorphism (PCR-ACRS)

This is more advanced and novel technique used for genotyping. PCR-RFLP becomes invalid for the alleles of genes where a naturally occurring restriction digestion site is not available. In that context, scientists have devised these molecular biology techniques, where a mutation in one of the primers is created so that a restriction digestion is induced in one allele of a gene only, the other form remains unchanged. On agarose gel the same pattern is obtained as in the PCR RFLP. The homozygous allele without restriction site development yields a single band, homozygous allele with an induction of restriction digestion yields two bands (large and a small DNA fragments) and the heterozygous form yields three bands.

During my PhD programme at NDRI, in 2012, we employed this technique for the screening of animals (Sahiwal, Tharparkar, Karan Swiss and Karan Fries) for A1 and A2 alleles of β-casein of milk protein. Similarly other Investigators (Kaminski and Olenski) used PCR-ACRS for genotyping of Czech Jersey and Holstein Friesian cattle for A1 and A2 alleles of β-casein of milk protein.

1.4.3.6 Allele Specific-Polymerase Chain Reaction (AS-PCR)

A variation of PCR exploited for screening purpose that directly detects a point mutation in organisms by examining the products of PCR. These products can be visualized in agarose or polyacrylamide gel already stained with ethidium bromide. This technique works on the principle that one of the primer has a mismatch at the 3′ end for one of the template that will not be allowed to extend and amplify. The primers without a mismatch at 3′ end will be amplified and employed for detection of the alleles of unknown genotype.

Ganguly in 2013 used this unique molecular biology technique for screening of cattle (Ongole, Frieswal Heifers, and Frieswal Bulls) for A1 and A2 alleles of β-casein of milk protein.

In this chapter, cow milk protein, its composition and nutritional value is explained comprehensively. It provides biomolecules like carbohydrates (lactose), proteins (casein and whey proteins), lipids, cholesterol, vitamins and minerals. Moreover, considerable stress is given on the β-casein variant of cow milk and its polymorphism. A1 and A2 alleles of this protein are elucidated completely as these are the key alleles for the foundation of A1/A2 milk protein hypothesis. Lastly, the techniques employed for the screening of such alleles (A1/A2) are described comprehensively. In the next chapter, I will discuss the A1/A2 milk protein hypothesis, its foundation and the evidences in favour and against this hypothesis.

References

Aleandri R, Buttazzoni L, Paggi U, Davoli R, Nanni Costa L, Russo V (1997) Effect of bovine milk protein polymorphism at two loci on cheese producing ability. International Dairy Federation special issue, pp 414–422

Aschaffenburg R (1961) Inherited casein variants in cow's milk. Nature 192(4801):431–432

Aschaffenburg R (1963) Inherited casein variants in cow's milk. II. Breed differences in the occurrence of beta-casein variants. J Dairy Res 30:251

Aschaffenburg R (1968) Genetic variants of milk proteins: their breed distribution. J Dairy Res 35(3):447–460

Baranyi M, Bosze ZS, Buchberger J, Krause I (1993) Genetic polymorphism of milk proteins in Hungarian spotted and Hungarian grey cattle: a possible new genetic variant of beta-lactoglobulin. J Dairy Sci 76(2):630–636

Bech AM, Kristiansen KR (1990) Milk protein polymorphism in Danish dairy cattle and the influence of genetic variants on milk yield. J Dairy Res 57(1):53–62

Boettcher PJ, Caroli A, Stella A, Chessa S, Budelli E, Canavesi F, Ghiroldi S, Pagnacco G (2004) Effects of casein haplotypes on milk production traits in Italian Holstein and Brown Swiss cattle. J Dairy Sci 87(12):4311–4317

Caldwell J, Weseli DF, Cartwright TC (1971) Occurrence of αs1-and β-casein types in five breeds of beef cattle. J Anim Sci 32(4):601

Cieślińska A, Fiedorowicz E, Zwierzchowski G et al (2019) Genetic polymorphism of β-casein gene in polish red cattle-preliminary study of A1 and A2 frequency in genetic conservation herd. Animals (Basel) 9(6):E377. https://doi.org/10.3390/ani9060377

Dar AH, Kumar S, Singh DV et al (2018) Distribution of allelic and genotyping frequency of a1/a2 allele of beta-casein in Badri cattle. Int J Livest Prod 8(2):1–10

Dhanammal CLM (2006) Studies on DNA polymorphism of αs1 and β-casein genes in Sahiwal cattle. Ph.D. Thesis. National Dairy Research Institute (Deemed University), Karnal, India

FAO Publications (2019a) Milk and milk products. In: Food outlook

FAO Publications (2019b) Dairy market review

Farkye NY (2003) Other enzymes. In: Fox PF, McSweeney PLH (eds) Proteins, Advanced Dairy Chemistry, vol 1, 3rd edn. Kluwer Academic/Plenum Publ., New York

Flynn A, Cashman K (1997) Nutritional aspects of minerals in bovine and human milks. In: Fox PF (ed) Lactose, water, salts and vitamins, Advanced Dairy Chemistry, vol 3, 2nd edn. Chapman & Hall, London

Fox PFA (1995) Lactose, water, salts and vitamins, Advanced Dairy Chemistry, vol 3, 2nd edn. Chapman & Hall, New York

Ganguly I, Kumar S, Gaur GK et al (2013) Beta-casein (CSN2) polymorphism in Ongole (Indian zebu) and Frieswal (HF x Sahiwal crossbred) cattle. Indian J Biotechnol 12(2):195–198

Groves ML (1969) Some minor components of casein and other phosphoproteins in milk. J Dairy Sci 52(8):1155–1165

Hallén E, Wedholm A, Andrén A, Lundén A (2008) Effect of β-casein, κ-casein and β- lactoglobulin genotypes on concentration of milk protein variants. J Anim Breed Genet 125(2):119–129

Jost R (2002) "Milk and dairy products" Ullmann's encyclopedia of industrial chemistry. Wiley, Weinheim. https://doi.org/10.1002/14356007.a16_589.pub3

Kaminski S, Cieoelinska A, Kostyra E (2007) Polymorphism of bovine beta-casein and its potential effect on human health. J Appl Genet 48(3):189–198

Kumar S, Singh RV, Chauhan A (2019) Molecular characterization of a1/a2 beta-casein alleles in Vrindavani crossbred and Sahiwal cattle. Indian J Anim Res 53:151–155

Lien S, Kantanen J, Olsaker I, Holm LE, Eythorsdottir E, Sandberg K, Dalsgard B, Adalsteinsson S (1999) Comparison of milk protein allele frequencies in Nordic cattle breeds. Anim Genet 30(2):85

Lin CY, McAllister AJ (1986) Effects of milk protein loci on first lactation production in dairy cattle. J Dairy Sci 69(3):704

McLean DM, Graham ER, Ponzoni RW, McKenzie HA (1984) Effects of milk protein genetic variants on milk yield and composition. J Dairy Res 51(4):531–546

Mishra BP, Mukesh M, Prakash B, Sodhi M, Kapila R, Kishore A, Kataria RR, Joshi BK, Bhasin V, Rasool TJ, Bujarbaruah KM (2009) Status of milk protein, β-casein variants among Indian milch animals. Indian J Anim Sci 79(7):722–725

Ng-Kwai-Hang KF, Kim S (1994) Genetic polymorphism of milk proteins in Ayrshire, Jersey, Brown Swiss and Canadienne. Brief communications of the XXIV International dairy Congress, Melbourne Gb74

Öste R, Jägerstad M, Anderson I (1997) Vitamins in milk and milk products. In: Fox PF (ed) Lactose, water, salts and vitamins, Advanced Dairy Chemistry, vol 3, 2nd edn. Chapman & Hall, London

Peterson RF, Kopfler FC (1966) Detection of new types of β-casein by polyacrylamide gel electrophoresis at acid pH: a proposed nomenclature. Biochem Biophys Res Commun 22(4):388–392

Raies MH (2013) Studies on release of β-casomorphins from milk and their effect on gut immunity. PhD thesis, NDRI, Karnal, Haryana. KrishiKosh. An institutional repository of Indian National Agricultural Research System

Raies MH, Kapila R, Shandilya UK et al (2012a) Impact of milk derived β-casomorphins on physiological functions and trends in research. Int J Food Prop 17(8):1726–1741

Raies MH, Kapila R, Shandilya U et al (2012b) Detection of A1 and A2 genetic variants of β-casein in Indian crossbred cattle by PCR-ACRS. Milchwissenschaft 67(4):396–398

Ramesha KP, Akhila R, Basavaraju M et al (2016) Genetic variants of β-casein in cattle and buffalo breeding bulls in Karnataka state of India. Indian J Biotechnol 15(2):178–181

Roginski H (2003) Encyclopedia of dairy sciences. Academic, London

Saran M (2017) A1 and A2 polymorphism in β-casein gene and lactoferrin gene fragment in indigenous cattle breeds. M.V.Sc. Thesis, Rajasthan University of Veterinary and Animal Sciences, Bikaner

Chapter 2
A1/A2 Milk Hypothesis

Abstract The generated ecological data, human case-control findings, in vivo and in vitro studies followed by biochemical and pharmacological mechanisms are strong enough to establish a correlation between A1 "like" milk consumption and increased risk for various health disorders. BCM-7 released from A1 milk is actually considered as a proposed risk element. However, the counter arguments furnished by EFSA, American Nutritionists and Truswell can't be ignored. The present status of the hypothesis isn't in a position to make consumer health recommendations. More mechanistic studies involving well-designed animal and human trials with explored cellular, molecular and immunological mechanisms are needed to reach at final conclusions.

2.1 Foundation of A1/A2 Milk Hypothesis

During early 90s, a surprisingly potential A1/A2 cow milk hypothesis was established by Elliot, McLachlan and coworkers. They proposed that a factor from the protein fraction of cow milk has a considerable role to play in the incidence of type 1 diabetes mellitus (T1D), coronary heart disease (CHD), schizophrenia and autism (McLachlan 2001; Laugesen and Elliott 2003; Sun et al. 2003). This astonishingly important protein was the A1 "like" β-casein protein. As in the aforementioned chapter discussed comprehensively, β-casein protein of cow milk has 12 genetic variants which are divided into two categories i.e., A1 and A2 "like" β-casein. According to this hypothesis, the A1 "like" components correlated with the incidence of these health complications. Further protein analysis established the presence of histidine and proline at position 67 in A1 and A2 "like" variants of β-casein protein respectively. The differential amino acid sequence in the two types of proteins in addition to the enzymatic cleavage patterns showed that a 7 amino acid peptide is released from only A1 "like" variants of β-casein of cow milk. This peptide is released from β-casein of milk protein and was found to have "morphine" like activity, therefore named as β-casomorpin-7 (BCM-7). This peptide was actually correlated with the incidence of these disorders. There are many ecological, case control reports, few animal and in vitro studies available that correlates this

M. R. Ul Haq, *β-Casomorphins*, https://doi.org/10.1007/978-981-15-3457-7_2

protein or peptide with these health complications. On the other hand, a report published by European Food Safety Authority in 2009 stated that there aren't adequate evidences available to draw final conclusions and recommendations in this perspective. The authority apprehends earlier data as mere suggestive evidence and refutes these well-established cause-effect correlations (EFSA 2009).

The literature available in favour of the hypothesis, the critical analysis against the same (data or commentaries) and besides, my personal comprehension and expertise in the same field for about a decade enables me to state that the data is enough for the establishment of this hypothesis. However, more mechanistic studies involving well-designed animal and human trials are required to confirm the hypothesis. Moreover, well designed cellular, molecular, biochemical and immunological parameters must be analysed to reach at final conclusion. The determination of signal cascade mechanisms in vitro in cell lines with more advanced biochemical and molecular techniques will unravel the hypothesis.

Overall, the hypothesis is fascinating and possibly significant, however, for further confirmations, verified and authenticated research with reproducible results is needed to make final public health recommendations and guidelines regarding world's dairy sector.

2.2 Genetics of A1/A2 Milk Hypothesis

The β-casein protein of cow milk has 209 amino acids. It has at least 12 genetic variants with variations at different amino acid positions. These variants are represented as A1, A2, A3, B, C, D, E, F, G, H1, H2 and I. The most frequently occurring alleles are A1 and A2, these variants are significant from A1/A2 hypothesis point of view. These alleles are located on exon 7 of chromosome no. 6. These two alleles show co-dominant expression, therefore, the genotypes can exist in homozygous (A1A1, A2A2) or heterozygous (A1A2) form. A1 allele frequency is mostly found in the cows located in the regions of North Europe i.e., Ayrshire, Shorthorn Friesian, British and Holstein. The A2 allele occurs more frequently in the cows located in the Channel Island (Jersey and Guernsey), Limousin and Charolais and Southern French breeds. More importantly, India the largest producers of milk worldwide, inhabits both the indigenous and crossbreed cattle. The indigenous cows (Sahiwal, Tharparkar, Ongole, and Vrindavani) contain milk predominantly of A2 type. However, the crossbreed cattle (Frieswal Heifers, Frieswal Bulls, Karan Fries and Karan Swiss) possess both A1 and A2 alleles, with latter prevailing the former allele. Fortunately enough the milk of Indian ingenious cattle is predominantly A2 while in crossbred cattle both variants occur with varying frequency like, Frieswal Heifers, Frieswal Bulls, Karan fries, Karan Swiss, Malnad (Gidda), Kasargod, Vrindavani and buffalo (Murrah and River) (Raies et al. 2012; Ganguly et al. 2013; Ramesha et al. 2016; Kumar et al. 2019). The molecular biology testing techniques revealed the A2 type alleles in Indian cows, the favorable allele of β-casein. Therefore, it indicates the Indian cow milk of premium or superior quality with flourishing global market (if hypothesis proven correct with deeper research). The favoured variant can be increased in the cow population through appropriate genetic breeding strategies. These comprises of the indirect method of picking A2 allele of bull semen or through more active method that needs screening the dairy animals for these alleles and then careful removing of A1 calves and retaining of A2.

2.3   Scientific Evidence Behind A1 Milk Hypothesis

The scientific literature favoring the A1/A2 hypothesis stands on several factors. The following factors are represented in Fig. 2.1.

- Ecological correlations
- Animal trials
- Case controls studies
- Biochemsitry (β-casomorphin release from A1 β-casein)
- Pharmacology (opioid nature of these peptides)
- Subjective experiences
- Milk Allergies
- Milk Intolerances

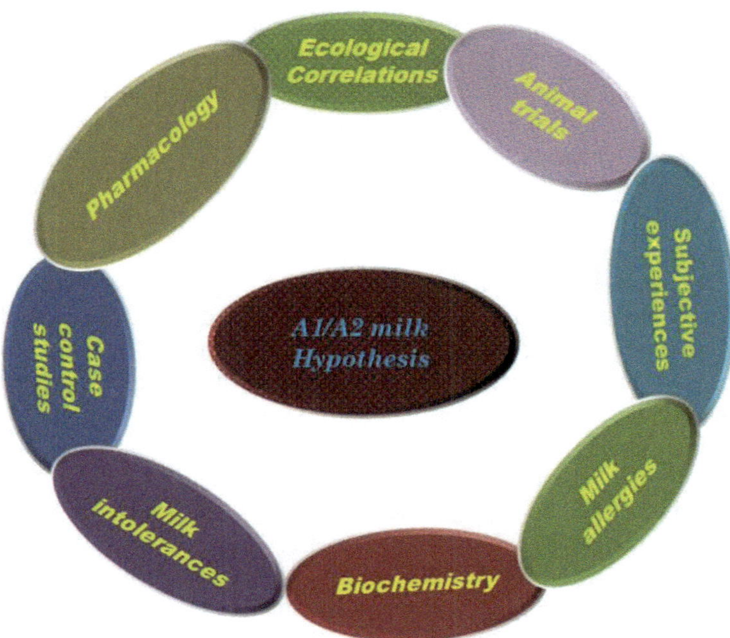

Fig. 2.1   Evidences in favour of A1/A2 cow milk hypothesis

2.3.1 Ecological Correlations

Ecological studies have linked consumption of A1 milk with T1D, cardiovascular diseases (CVD) and sudden infant death syndrome (SIDS). Elliot and coworkers in 1999 compared consumption of A1 "like" milk with consumption of T1D. A correlation was established between consumption of cow milk protein taken annually at national level with the incidence of T1D in children aged 0–14 years. The composition of the breeds of cows and milk protein polymorphism was kept under consideration. Moreover, those countries were chosen for studies that had very low dairy inputs from other countries. They found that the total protein intake didn't correlate ($r = +0.402$) with incidence of T1D. However, to the surprise, the intake of A1 β-casein correlated well ($r = +0.726$) with the incidence of T1D. This effect was found more pronounced ($r = +0.982$) on consumption of A1 "like" β-casein (A1 + B). The B variant having histidine at 67th position of this protein is included in A1 "like" milk. The researcher therefore concluded that A1 "like" variants on gastrointestinal digestion release is 7-amino acid peptide, however, A2 "like" variants are unable for such release. Since this peptide is found to have opioid property including immunosuppressive activity, that may explain the specificity of the correlation between the intake of A1 "like" but not A2 "like" β-casein variants and T1D (Elliot et al. 1999). Similarly, Thorsdottir and coworkers in 2002 found low proportion of A1 "like" cow milk (A1 + B) in Icelandic compared to Scandinavian countries. They couldn't establish a correlation between cow milk consumption and incidence of T1D in the infants of Iceland. The investigators conclude that the lower contents of A1 "milk" in these countries could account for lower incidence of T1D in Iceland in comparison with Scandinavia (Thorsdottir et al. 2002). In 2001, McLachlan conducted a study in children aged below 15 year in 16 countries. The data generated established a strong correlation ($r^2 = 0.73$) between A1 milk intake and incidence of ischaemic heat disease. However, the consumption of total milk protein intake correlated weakly ($r = 0.23$) with incidence of T1D (McLachlan 2001). The aforementioned correlation was supported incidence of T1D in Nordic Countries where higher incidence was observed compared to Iceland. The dairy cows of the former countries produce more A1 'like" milk, therefore, could account for this strong correlation (Birgisdottir et al. 2002). Similarly, Laugesen and Elliott in 2003 developed a strong correlation ($r = 0.92$) with consumption of A1 milk and incidence of T1D (Laugesen and Elliott 2003). The factors (risk) responsible for many human illnesses and the data (mortality) was collected from World Health Organization (WHO). Both the protein consumption data of milk and nutritional data were taken from Food and Agriculture Organization (FAO). The fractions of β-casein and data of food supply were estimated by breed from the literature of dairy science. The major cardiovascular diseases in humans include coronary or ischemic heart disease (CHD or IHD). According to A1 milk hypothesis, the consumption of A1 milk is correlated well with the incidence of CHD. The study was performed in adults aged between 30–69-years males. The data was taken across 16 countries including Canada, Australia, Denmark, Austria, Finland, Iceland, France, Israel, New Zealand, Japan, Norway, Sweden, Scotland, United Kingdom, West

Germany, and USA. The correlation was established between milk protein intake and mortality rate due to IHD. The results indicated a strong correlation ($r = 0.86$) between A1 milk intake and CHD. The intake of A1 β-casein was also related with red meat and animal fats (general risk agents) and conventional risk agents like hypertensives, smokers, body mass index and blood cholesterol levels. It was found that except β-casein variants, other risk agents couldn't correlate well with the incidence of CHD. The investigator finally concluded that A1 β-casein or its derived fragment BCM-7 may be a functional element for the contribution towards the incidence of cardiovascular disease. These studies were further strengthen with the ecological findings of Laugesen and Elliott (2003) who developed a correlation between consumption of A1 β-casein and mortality rate due to CHD. This study was performed in 35–64-year-old males and involved 19 countries including Canada, Australia, Finland, Austria, Denmark, France, Germany, Iceland, Hungary, Israel, Japan, Italy, New Zealand, Sweden, Norway, Switzerland, United Kingdom, Venezuela, USA). It was found that consumption of A1 milk has a strong correlation ($r = 0.76$) with CHD compared to total protein ($r = 0.60$), fat and fat derived dietary compounds (Laugesen and Elliott 2003).

SIDS is a cause of death in infants up to 1 year. Quite evidently, the sole source of food for the infants is milk, therefore, one of the factors common to all infants that acquire SIDS is milk. A well-established fact is that children's gastrointestinal tract and central nervous system is immature, therefore, the generated peptides from milk can cross the gut that is not fully developed and bears enough chances for the small peptides to cross blood-brain barrier. In SIDS the children with abnormal respiratory control and vagal nerve development, the BCM-7 peptides having opioid properties may lead to depression of the brain-stem respiratory centers, leading to death (Sun et al. 2003). Reports have also detected the presence of BCM immunoreactive material in the human children brain stems. Further studies have clearly demonstrated that BCM-7 (small peptide) and the immature gut of infants provides favorable environment for the peptide to absorb through gut cross blood brain barrier and bind various opioid receptors in the nervous, endocrine, and immune systems (Bell et al. 2006). The transport of BCM peptides across central nervous system of mice and rats has also been established (Sun et al. 2003).

Some epidemiological reports have linked consumption of A1 β-casein with neurological disorders like autism and schizophrenia. The elevated levels of BCM-7 have been detected in urine and blood of these patients (Sun and Cade 2003; Reichelt and Knivsberg 2003). These investigators also hypothesized that these peptides crosses the gut and reaches the blood in these patients, crosses the blood-brain barrier, and affects the neurological processes (Sun et al. 2003).

2.3.2 Animal Trials

The role of casein in general and β-casein in particular in human diseases was investigated in the early 80s. Elliot and Martin in 1984 performed a trial in Biobreed rats (BB) and established a correlation between protein intake (milk) and incidence of diabetes (Elliot and Martin 1984). BCM-7 as a diabetogenic agent from A1 milk was further supported with the observation of in a report regarding β-casein protein of cow milk as "Jeckyl and Hyde" (Elliott et al. 1996). An investigation was carried out in non-obese diabetic mouse, demonstrating the role of different diet compositions on diabetes incidence rates (Elliott et al. 1997). They reported a considerable effect on diabetes incidence with diet formulations added with A1 (47%) compared to A2 β-casein (0%). Results also indicated a remarkable decrease in diabetes incidence from 28% to 2% on administration of casein hydrolysate. This diabetogenic effect incidence with A1 β-casein was improved on naloxone (antagonist of BCM-7) administration in mice models. These observations, therefore, evidently show A1 β-casein as a promoter and A2 β-casein a protector against diabetes incidence. However, the critical analysis reveals that this animal trial had a lacuna in terms of not revealing the diet formulations. This trial was followed by animal trials performed in New Zealand, UK and Canada. These countries are good in milk production and demonstrate a remarkable potential for milk and other milk products. There were inconsistent results in the three countries, the trials performed in New Zealand didn't end with final conclusions, as these animals got infected and died. The Canadian trials as expected established that A1 β-casein slightly more diabetogenic compared to A2 β-casein in mice. However, to a surprise, the UK trails showed something were different, regarded both types of diets containing A1 and A2 β-casein protective against diabetes incidence (Beales et al. 2002). As a part of discussion, it may be presumed that the unpredicted results obtained in the aforementioned trials may be due to following reasons:

- The full composition of all the diets was not revealed
- Trials used diet formulations containing wheat that contains gluten, the latter is a source of gluteomorphins that show opioid properties like that of casomorphins and could account for the diabetogenicity incidence.
- Pregestimil, the infant formula used in the trials could be a source of BCM-7 (Cone 2002; Australian Broadcasting Corporation 2003).

Keeping in view the above discussion, it may be put forward that there is lack of perfect 'control' that may account for inconsistencies in the results of the above trials. The presence of peptide (BCM-7) and supplementary diabetogenic representatives resulting from wheat may have reproduced in the results of these animal trials. Therefore, keeping these voids under consideration, the results obtained may be misleading till published diet compositions with basal diet negated with wheat is used in the perfect animal trials. In this connection, the researchers or agencies that are against the hypothesis making Beals's observations as the standard must further anticipate that the animal trials were sponsored by the Fonterra group and the diet formulations were given by New Zealand Dairy Research (part of Fonterra). Both the agencies were advocating against this hypothesis. In this aspect, in 2018, a group of investigators from Australia analyzed the role of A1 β-casein in cow milk on incidence of T1D in non-obese diabetic mice. They observed that the early life exposure to milk rich in A1 β-casein has role in incidence of T1D and that is generation dependent. They further reported subclinical insulitis and changed glucose management with a significant decline in proportion of non-conventional regulatory T cell subset in terms of $CD4^+ CD25^- FoxP3^+$. Consequently, they concluded that consumption of A1 β-casein alters the glucose homeostasis and encourages the incidence of T1D (Chia et al. 2018). On the contrary, a group of researchers from China observed the protective role of BCM-7 against hyperglycemia and free radical-mediated oxidative stress in rats (Yin et al. 2010, 2012). There were two lacunas in these trials, the experiments were carried out with commercially synthesized BCM-7 instead of native A1/A2 β-casein proteins. Moreover, these experiments weren't conducted in diabetic mouse models (NOD mice) rather diabetes was induced with streptozotocin. In conclusion, there is enough literature available that correlates consumption of A1 milk with incidence of T1D. The controversial reports promoting against this hypothesis lacks perfect diabetic animal models, the diet formulations are fully or partially unknown or reports are just commentaries. Nevertheless, the arguments and counter-arguments demand more research with consistent and reproducible result to make final public recommendations.

In 2014, we also took an animal trial in mice, analyzed the effect of feeding three variants of β-casein (A1A1, A1A2 and A2A2) on the gut function, the latter was taken for the study due to the fact that the gut is the first organ of the immunological interactions for molecules taken orally. The animals were classified into four groups fed with control basal diet and three other groups were fed with basal diet replaced with protein fraction of A1A1, A1A2 and A2A2 β-casein variants. The gut function was studied through measurement of myeloperoxidase activity (MPO), inflammatory markers (MCP-1and IL-4), total antibodies (total IgE, IgG, sIgA, IgG1 and IgG2a) and expression studies of toll-like receptors (TLR-2 and TLR-4). The histological studies of goblet cells, total leukocytes and IgA were also done. The results showed that oral consumption of homozygous (A1A1) heterozygous (A1A2) (also regarded as A1 "like" variants) remarkably increased the levels of myeloperoxidase, monocyte chemotactic protein-1, interleukin-4, total IgE, IgG, IgG1, IgG2a and infiltration of leukocytes in intestine. Further expression studies depicted an increase in TLR-2 and TLR-4 mRNA expression on oral consumption of A1 "like" variants. Nevertheless, there were non-significant changes observed in sIgA, IgA^+ and goblet cell on oral intake of β-casein variants. Therefore, we concluded that oral intake of A1 "like" variants of β-casein induced inflammatory reactions in the intestine through activation of Th2 pathway in comparison to A2A2 variants. Our study therefore agreed with the previous findings of the adverse effect of A1 "like" milk and proposes potential exacerbation of inflammatory reaction for etiology of various health complications (Raies et al. 2014a).

In the same year, we performed another animal trial in mice to establish the role of commercially synthesized peptides BCM-7 and BCM-5 on gut immune function. These peptides were selected with the reason that these are released from A1 "like" variants of β-casein and regarding the ill effects of this casein. These peptides were given orally to cross check the effects of this β-casein. These peptides were reconstituted in water and given orally to mice at a dose of 7.5×10^{-8} mol/day/animal. The trial lasted for 15 days. The results indicated that oral consumption of BCM-7 and BCM-5 increased the levels of inflammatory molecules like myeloperoxidase, monocyte chemotactic protein-1, interleukin-4 and allergic molecule, histamine. The humoral immune response (IgE, IgG, IgG1/IgG2a), and leukocyte infiltration also increased in intestine. The expression studies indicated an increase in the mRNA levels of toll like receptors including TLR-2 and TLR-4 in intestine. Expectedly, there were non-significant changes in sIgA, IgA$^+$ and goblet cell numbers. In conclusion, the results of this study showed that oral intake of BCM-7/5 peptides induced inflammatory reaction response in intestine and the effect was most probably through Th2 pathway. These results confirmed our previous results obtained on intact protein and validated the inflammatory response of A1 "like" milk (Raies et al. 2014b).

2.3.3   Case Control Studies

Cavallo and coworkers in 1996 reported an increased antibody profile against β-casein of cow milk protein irrespective of the genetic variants of β-casein (A1/A2) in T1D (Cavallo et al. 1996). T cells that were sensitized on intake of β-casein in T1D have enough chances to cross react with β-cells of islets of langerhans located in pancreas. However, two conditions couldn't be justified that include β-casein protein polymorphism and explanation of a cause-effect relationship (Monetini et al. 2002). Patients with T1D ($n = 287$), their siblings ($n = 386$), parents ($n = 477$) and healthy controls ($n = 107$) were checked for antibody response against A1 and A2 β-casein milk proteins. They found elevated antibody response against A1 β-casein of milk protein in patients (diabetic) and their siblings. Additionally, antibody profile was elevated against β-casein and a significant overlap persisted with controls in addition to an enhanced antibody response against other antigens also (Padberg et al. 1999). In this perspective, this is confusing to note whether this immune response is linked with etiology of disease (diabetes) or is a manifestation of illness. Moreover, longitudinal trials and human feeding are significantly desirable to fully understand the pathogenesis of this disorder (diabetes) on consumption of A1/A2 β-casein milk proteins. These trials are possible in genetically at-risk infants and could essentially contribute to the evidence base.

2.3.4   Biochemsitry (β-Casomorphin Release from A1 β-Casein)

There are many biochemical hydrolysis protocols representing the simulated gastro-intestinal digestion (SGID) that show the two variants of β-casein are hydrolyzed differently using different combination of enzymes. Jinsmaa and Yoshikawa used these enzymes and detected the release of BCM-7 from only A1 "like" varaints. During my PhD programme at National Dairy Research Institute (NDRI), we also checked the release of BMC-7 and BCM-5 from A1A1, A1A2 and A2A2 genetic variants of β-casein. This release was assessed after SGID using enzymes pepsin, trypsin, chymotrypsin, elastase and leucine amino peptidase. The three variants of β-casein were isolated from milk taken from crossbred cattle Karan Fries. These cattle were first screened for these genotypes using polymerase chain reaction-Amplification Creation restriction Digestion (PCR-ACRS). The detection of the peptides was carried out with analytical HPLC and the fractions were separated using preparative HPLC. The peptides were quantified with competitive ELISA and sequences of the peptides were elucidated with MS-MS spectrometry. Among many fractions collected, the fraction B showed the presence of a peptide sequence with 14 amino acids (VYPFPGPIHNSLPQ). This 14 amino acid peptide has in its sequence encrypted as an internal BCM sequence. When this 14 amino acid sequence was further digested with elastase and leucine amino peptidase, it revealed the presence of BCM-7 that was quantified with competitive ELISA as shown in Fig. 2.2. The amount of BCM-7 released from A1A1 was found to be 0.20 ± 0.02 mg/g β-casein that was almost 3.2 times more than heterozygous A1A2 variant of β-casein. On the other hand, as expected, the release of BCM-7 was not observed from A2A2 β-casein variant. Moreover, the release of BCM-5 was not detected from any of the variants of β-casein (Raies et al. 2015).

In 1999, Jinsmaa and Yoshikawa have observed that pepsin and leucine amino-peptidase release the amino end of the BCM-7, the pepsin hydrolysed the peptide bond between leucine and valine located at position 59 and 60 respectively, and then leucine aminopeptidase removed the valine from the amino terminal. The pancreatin elastase was responsible for creating the carboxyl terminus hydrolyzing the bond between isoleucine and histidine located at position 67 and 68 respectively. Trypsinization produced large peptides from the proteins having specificity for basic amino acids. Chymotrypsin also released large fragments from the protein cleaving the peptide bond at the aromatic amino acids. In our study, we also found that BCM-7 is released from the homozygous A1A1 and heterozygous A1A2 β-casein variants by pepsin, trypsin and chymotrypsin when subsequent hydrolysis was carried out with elastase and LAP (Jinsmaa and Yoshikawa 1999). De Noni in 2008 quantified the release of BCM-7 (0.136–4.7 mg/g β-casein) from A1 "like" variants of cow β-casein (Holstein-Friesian and Jersey) (De Noni 2008). These observations support our data obtained in Karan Fries (Indian crossbred). However, we obtained a lower yield (0.200 ± 0.02 mg/g β-casein) compared to that quantified in Holstein-Friesian and Jersey. This difference in the BCM-7 may be multifactorial including the hydrolysis protocol, the genetic makeup and phenotypic expression of the protein (β-casein) of the investigated cows. They executed SGID protocols using pepsin and pH adjusted to 2.0, 3.0 and 4.0 that was subsequently hydrolyzed with Corolase PP™ in Jersey and Holstein-Friesian breeds. The yield of BCM-7 from A1A2 β-casein variant (heterozygous) using PeTEL or PeTCEL enzyme combinations respectively was revealed in our study. The production of BCM-7 was found to be lesser by about 3.2 fold compared to that obtained from homozygous A1A1 β-casein. The results of our study were in agreement with the studies of De Noni (2008) who detected a lesser production of BCM-7 (0.005–0.435 mg/g β-casein)

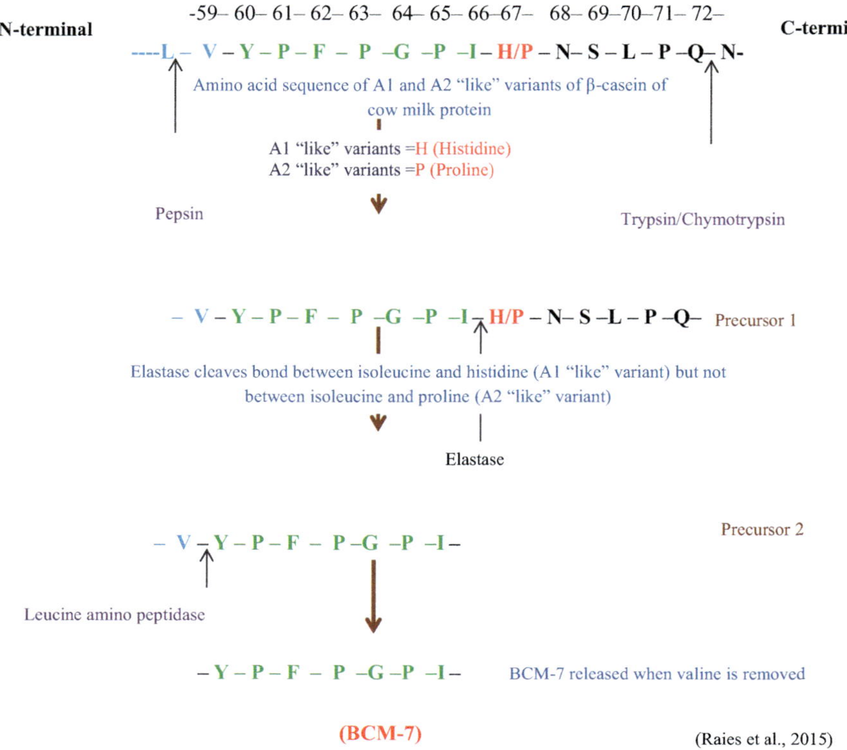

Fig. 2.2 Release of BCM-7 from A1 "like" variants of β-casein of cow milk by simulated gastro-intestinal digestion

from A1A2 β-casein variant (heterozygous) compared to that in A1A1 (0.136–4.7 mg/g β-casein) β-casein variant of cow milk.

The lesser BCM-7 production from heterozygous proteins is thought to be the lesser amount of A1 variant in A1A2 variants compared to homozygous A1A1 variants. These alleles demonstrate co-dominant expression and variability in the levels of gene expression which exemplifies the variants of β-caseins (Bleck et al. 1996). These variants of β-casein might affect the protein synthesis mediated either by decreased mRNA stability or various expression rates (Rando et al. 1998). Contrariwise to the above results, to the best of our knowledge, there is only one report available that demonstrate the release of BCM-7 from the homozygous A1A1 variant of β-casein. Nevertheless, the amount of the BCM-7 released from A1A1 variant (11.59 mg/g of extract) was found almost four (4) times higher compared to that yielded from A2A2 homozygous A2A2 β-casein variant (2.87 mg/g of extract). The observations of this research group seem to be little biased due to many reasons. They performed SGID with pepsin for 24 h, this hydrolysis time contradicts with actual hydrolysis incubation time of SGID (2–5 h). The other reason is the method of detection of peptides, they used only HPLC/UV for analysis including both detection and quantification, this method alone is not effect for analysis as one peak comprises of many co-eluting peptides having same hydrophobicity and in turn same retention peaks. Therefore, this method should always be followed by ELISA and MS–MS instead of HPLC alone. In our study, we couldn't find any release of BCM-5 form any of the varaints of β-casein of cow milk with SGID. Therefore, our results are in agreement with earlier studies of various investigators (De Noni 2008; De Noni and Cattaneo 2010; Svedberg et al. 1985). These researchers found the release of BCM-7 but not BCM-5 from milk and other dairy products including commercial cheeses, milk-based infant formulas, probiotic fermented milks, yoghurts, experimental infant formulas and dried milk derivatives. The reason for this non-production of BCM-5 from BCM-7 seem to the high content of proline (43%) in the latter and this amino acid is really very resistant to hydrolysis (Kim et al. 1972), therefore, the production of BCM-5 was unlikely. In conclusion, our study in association with earlier data for release of BCM-7 from A1 "like" varaints of β-casein from cow milk confers the fundamental information for further studies regarding the potential bioactivity of BCMs.

Cieslinska and coworkers in 2007 compared the release of BCM-7 in milk (fresh) and pepsin hydrolysed milk. They assessed a higher yield of BCM-7 in hydrolyzed milk composed of A1 component of β-casein compared to the hydrolyzed milk with A2 variant of β-casein. The yield was almost four (4) times higher in former compared to latter. Further, the yield was very low in fresh milk (Cieslinska et al. 2007). The significance of the De Noni's work in 2008 was the use of three varaints of β-casein at different pH (2.0, 3.0 and 4.0) followed by Corolase PPTM, a mixture of pancreatin enzymes using SGID. They estimated a higher yield from B (176 mmol/mol β-casein) variant which is one of the variant of A "like" milk followed by A1 β-casein (20 mmol/mol β-casein). Nevertheless, they couldn't find any release from A2 variant of β-casein during the whole process of SGID (De Noni 2008). The same research group extended this research work and in 2010, they detected the release of BCM-7 in unprocessed milk, whole ultra-heat treated milk, whole pasteurized milk, skimmed in-bottle-sterilized milk, fermented milks (yoghurt) and probiotic fermented milk. The carried out the hydrolysis protocol with the same SGID method using pepsin and Corolase PPTM. They estimated the highest yield in unprocessed milk (1.16 mg/mg) that was then followed by whole ultra-heat treated milk (0.96 mg/mg). The whole pasteurized milk depicted a higher release (0.64 mg/mg) of BCM-7 compared to skimmed in-bottle-sterilized milk (0.60 mg/mg). In yoghurt (fermented milks) the BCM-7 was released at a concentration of 0.65–0.98 mg/kg however, BCM-7 was detected in probiotic fermented milk (PFM) at a concentration of 0.29–1.23 mg/kg. The results also indicated the release of this peptide (BCM-7) in powdered milk including sodium caseinate (17.68 mg/kg) skim milk powder (3.46 mg/kg), and milk protein caseinates 1 and 2 (16.99–22.18 mg/kg). The babies who have less access to mother's milk have alternatives in the form of infant formulas. The latter have in their composition the alternative variants of β-casein of cow milk. Therefore, it is quite possible that these products may be source of these BCM-7 peptides. In this perspective, Hernandez-Ledesma and co-workers in 2004 screened these infant formulas for the release of BCM-7, hydrolyzing these products with pepsin at pH 3.5 which was then followed by hydrolysis with Corolase PPTM (Hernández-Ledesma et al. 2004). Unexpectedly, the research group couldn't detect any release of BCM-7 sequence from these products. However, BCM-9 was detected after SGID from these infant formulas (Hernández-Ledesma et al. 2007). Jarmolowska and its associates in 2007 worked on the infant formulas for the release of BCM-5 or BCM-7. They successfully assessed the release of BCM-5 from infant formulas using pepsin individually or in combination with other enzymes (pepsin and trypsin). They detected the release of BCM-5 at a concentration of 0.67 nmol/mL and 74.46 nmol/mL with pepsin and combination (pepsin and trypsin) hydrolysis respectively. Nevertheless, BCM-7 couldn't be determined from the infant formulas either previous or subsequent to SGID (Jarmolowska et al. 2007). The importance of SGID using all the simulated enzymes existed when De Noni was unable to assess either BCM-5 or BCM-7 only after pepsin hydrolysis of β-casein. The release of BCM-7 was possible only after using mixture of pancreatic enzymes (Corolase PPTM) and this peptide was detected at a concentration of 0.02–0.37 nmol/mL were detected (De Noni and Cattaneo 2010). Fermented dairy products are those milk products that are fermented with lactic acid bacteria like Lactococcus, Lactobacillus etc. These are also called as cultured dairy products. The release of BCMs from these cultured milk foods is unlikely; because these bacteria release dipeptidyl peptidase including X-prolyl-diaminopeptidyl peptidase (Law and Handrickman 1997), the latter have specificity for proline amino acid residues and therefore have enough chances to hydrolyze these peptides, as these fragments possess maximum of these amino acid residues. When in vitro trials were carried out using bacteria deficient in these enzymes (Matar and Goulet 1996), the release of these peptides was detected from these cultured dairy products (Haileselassie et al. 1999). For the first time in 1985, Hamel and his associates detected some molecules resembling BCMs and called BCM immunoreactive material (BCMIR) from cow milk when the latter was fermented with lactic acid bacteria (LAB) (Hamel et al. 1985). When chemically synthesized and commercially bought BCM-7 peptides were hydrolyzed with cell-free extract of PepX-deficient mutant strains of Lactobacillus helveticus L89 for 120 min, the release of BCM-4 was detected. Also it was reported that BCM-4 is released from pasteurized milk and cultured with PepX-deficient mutant of Lactobacillus helveticus L89 (Matar and Goulet 1996). When fermentation of skim milk was carried out with pasteurized milk and fermented with a PepX-deficient mutant of Lactobacillus helveticus L89, the release of BCM like peptides were detected in HPLC followed by mass spectroscopy (Schieber and Brückner 2000). Many reports have demonstrated production of BCM-11 and BCM-4 from heat treated milk at 142 °C for a short duration of 3–4 s using probiotic Lactobacillus GG strain followed by hydrolysis with pepsin and trypsin. These peptides were not released without pepsin and trypsin hydrolysis (Rokka and Syväoja 1997). Astonishingly, Kahala and associates couldn't detect any release of BCM from Finnish milk products (yoghurt) that were fermented with L. bulgaricus and S. thermophiles. The probable reason might be the source of A2 milk that is quite evidently not the source of these peptides BCMs (Kahala et al. 1993).

2.3.5 *Pharmacology (Opioid Nature of BCM Peptides)*

After the discovery of morphine (an opioid compound), there has been an increased curiosity for the scientists to discover new exorphins or exogenous opioid compounds and use them in human health, nutrition and disease. The more focus has been the components of diet including milk, wheat and other nutritional components. Among the many components of diet, theses opioid compound have been discovered from milk. From milk a class of peptides as aforementioned in the text, are BCMs. As the name indicates these peptides resemble "morphine", and are derived from β-casein milk protein and further look like "morphine" in structure and action.

The pharmacological studies have established that activity of BCM-7 similar to that of morphine and it narcotic or opioid properties were first demonstrated in 1979 from casein peptone. These findings were established in rats with BCM-7 and the effects counteracted by the use of naloxone, an opioid antagonist (Brantl et al. 1979).

The standard pharmacological assays that determine affinity and activity of opioid compounds including binding (ligand) assay, the mouse vas deferens (MVD) and guinea-pig ileum longitudinal muscle myenteric plexus preparation (GPI) have clearly established that bovine BCM-7 has ability to bind mu (μ)-opioid receptor. Moreover, they have been found to have some affinity towards delta δ-receptors. When activities were determined through GPI test, the bovine BCMs were found to be more potent (3–30 times) compare to human BCMs (Koch et al. 1985). The order of the potencies for the bovine BCMs include BCM5>BCM4>BCM-8>BCM-7. It was Brantl and coworkers in 1981 who assessed the binding affinities of the BCMs (bovine) for the mu (μ) opiate receptors and these receptors were taken from the tissue of rat brain, the values were indicated in terms of IC50 values and the results indicated that these peptides have lesser affinities (300 fold) for these receptors. Among BCMs analysed BCM-7 was found to have least binding affinity and BCM-5 to have highest affinity towards these μ-receptors of rat (Brantl et al. 1981).

2.3.6 *Subjective Experiences*

The subjective and experiential substantiation from consumers who complain of A1 milk intolerance like bloating, diarrhea, and nausea but tolerating A2 milk easily. The A1 "milk" science is highly controversial although very significant and influenced by commercial and intellectual property issues. The main companies or corporations advocating in this matter are A2 Milk Company and Fonterra. They have property rights for screening animals for A1/A2 alleles and market milk of different brands.

2.3.7   Milk Intolerances

As a part of smooth digestion, many milk consumers take A2 milk due to the fact that it is very easy for them to digest this type of milk. It is also crystal clear that the A2 milk comprises of lactose that is mostly related to intolerance in terms of lactose intolerance. However, the possible explanation for the conjugated effect of lactose and A1 milk/BCM-7 is having dual effect on intestinal tract.

The first one is that the BCM-7 produced from A1 milk in the gastrointestinal tract acts as an opioid peptide, therefore like other peptides, may have an effect on the gut that ultimately slows down the flow or passage of nutrients through the gut. The latter effect provides enough chances for the lactose to remain in the gut, provides enough time for fermentation of lactose in the tract. Quite evidently, the fermentation process demonstrates the exponential progress curve, therefore may produce more gases and other products.

The second explanation is direct i.e., many consumers of milk are directly intolerant to BCMs especially BCM-7. This effect can easily be crosschecked by giving the consumers (intolerant to ordinary cow milk) the goat milk, if they can digest latter, they can be friendly with A2 milk also.

2.3.8   Mild Allergies

Theoretically intolerances and allergies are quite dissimilar. The allergy is defined as "*an adverse health effect arising from a specific immune response that occurs reproducibly on exposure to a given food*" (Boyce et al. 2010). On the other hand, intolerances aren't related to immune reactions.

However, more often the two are associated with each other; the intolerance seemingly develops as a consequential reaction. The allergies (mild) that have been reported with the BCM-7 are asthma and eczema; however, the evidences are case-related. There are many findings available that demonstrate that administration of BCM-7 causes production of mucus (Claustre et al. 2002; Trompette et al. 2003; Zoghbi et al. 2006; Boyce et al. 2010), this is the most reasonable clarification for the consumers who take milk are associated with mucus production. There are many proteins in the milk unlike BCM-7 that may be responsible for adverse reactions that may include anaphylactic shock in consumers who are very susceptible to these shocks.

2.4   Commercialization of A2 Milk

The a2 Milk Company Limited was established by McLachlan and Paterson in 2000 in New Zealand. The headquarter is now established in Sydney, Australia. The company bears many investors at the international level. The company has its own property rights that allow screening of β-casein protein for their A1 and A2 alleles of cow milk. The chief objective of the company is to sell milk and other milk products including infant formulas, ice creams and powdered milk that are entirely composed of A2 milk. These milk and other dairy products are sold as the supreme or premium quality products. The company is also aimed at to establish the national and international cow and buffalo herds with A2 variant of β-casein. In past, the company has already launched milk and other milk products in many countries including New Zealand, UK, USA, China and Australia. (www.thea2milkcompany. com). Primarily, this initiative was intensely criticized by the leading milk company in New Zealand known as Fonterra. They were objecting the A1/A2 milk hypothesis. This was main hindrance in the progress of A2 Milk Company for delayed commerce at national and international level. The Fonterra did negotiate with a2 Milk Company through withdrawal of litigation issue in 2003. This was followed by licensing of a trademark and patent with Ideasphere Incorporated (ISI) and the main aim was to market A2 milk in North America in 2003 (www.thea2milkcompany. com). The company further progressed in 2007 and extended the deal with Original Foods Company to trade A2 milk in midwestern states (Burgess 2007). In 2011, the company took an undertaking with British Milk Supplier (Müller Wiseman Dairies) for selling a2 dairy products in Britain and Ireland (Teresa 2011).

Around 2013 to 2014, the a2 Platinum, a branded based Infant formula and Milk Thickened Cream both the dairy products made from A2 milk were launched in New Zealand, Australia and China (lynch 2015; Christopher 2013). Overall, Australia has denoted as an established and developed market for a2 Milk™ since 2007. This is the principal and supreme fresh milk product and easily found in supermarkets of Australia. It has a grocery price market share of around 10%. In the same country, the infant formula branded as a2 Platinum was launched in 2013 with 32% market share. The company is planning to launch new products like a2 Platinum premium stage 4 junior milk drink, Premium a2 Milk™ powder blended with Mānuka honey and a2 Platinum Premium Pregnancy milk formula in the financial year 2018. In New Zealand, the fresh milk brand, a2 Milk™ has been licensed to Fonterra and is expected to be available widely in New Zealand from October 2019. In People's Republic of China, the company is constructing a strong business due to wide brand awareness and Chinese consumers are encouraged to buy either the Australian a2 Platinum® or unique Chinese a2™. Currently, infant formula has achieved 5.1% consumption value share in china in addition to this fresh a2 Milk™ that is directly exported from Australia to china supermarkets. In US, the fresh milk a2 Milk™ was transported to supermarkets in 2015 and strong encouragements in addition to recognition as a leading milk innovator from highly regarded US press outlets including The Wall Street Journal, The Washington Post, CBS, and Forbes. The company is marked to focus on 6000 stores for A2 milk in the financial year 2018 with more attention to regions like California, Southeast and Northeast regions. The company is listed among the top ten most innovative companies in food in 2018 by US-based business magazine Fast Company. In UK, the a2 Milk Company stepped in UK in 2012, since then fresh milk brand a2 Milk™ increased from small base to in-store sales velocity. The company anticipates converting between traders of its fresh milk dairy product early in the economic year 2019 and change to a new package covering design.

References

Australian Broadcasting Corporation (2003) White mischief transcript. www.abc.net.au/4corners/content/2003/transcripts/s820943.htm

Beales P, Elliott R, Flohé S et al (2002) A multi-Centre, blinded international trial of the effect of A1 and A2 β-casein variants on diabetes incidence in two rodent models of spontaneous type i diabetes. Diabetologia 45(9):1240–1246

Bell JS, Grochoski TG, Clark JA (2006) Health implications of milk containing β-casein with A2 variant. Crit Rev Food Sci Nutr 46(1):93–100

Birgisdottir BE, Hill JP, Harris DP et al (2002) Variation in consumption of cow milk proteins and lower incidence of type 1 diabetes in Iceland vs the other 4 Nordic countries. Diabetes Nutr Metab Dis 15(4):240–245

Bleck GT, Conroy JC, Wheeler MB (1996) Polymorphisms in the bovine β-casein 50 flanking region. J Dairy Sci 79:347–349

Boyce JA, Assa'ad A, Burks AW et al (2010) Guidelines for the diagnosis and management of food allergy in the United States: report of the NIAID-sponsored expert panel. J Allergy Clin Immunol 126(6):S1–S58

Brantl V, Teschemacher H, Blasig J et al (1981) Opioid activities of beta-casomorphins. Life Sci 28(17):1903–1909

Brantl V, Teschemacher H, Henschen A et al (1979) Novel opioid peptides derived from casein (beta-casomorphins). Isolation from bovine casein peptone. Hoppe Seylers Z Physiol Chem 360(9):1211–1216

Burgess M (2007) For the New Zealand herald. A2 to tap into US milk market

Cavallo MG, Monetini L, Walker B et al (1996) Diabetes and cow's milk. Lancet 348(9042):1655

Chia JSJ, McRae JL, Enjapoori AK et al (2018) Dietary cows' milk protein A1 beta-casein increases the incidence of T1D in NOD mice. Nutrients 10:1291

Christopher A (2013) A2 gets ready to launch baby formula. The New Zealand Herald, Auckland

Cieslinska A, Kaminski S, Kostyra E et al (2007) β-Casomorphin-7 in raw and hydrolyzed milk derived from cows of alternative β-casein genotypes. Milchwissenschaft 62:125–127

Claustre J, Toumi F, Trompette A et al (2002) Effects of peptides derived from dietary proteins on mucus secretion in rat jejunum. Am J Physiol Gastrointest Liver Physiol 283(3):G521–G528

Cone DH (2002) Secret memo reveals Fonterra alarm National Business Review, p 14

De Noni I (2008) Release of beta-casomorphins 5 and 7 during simulated gastro-intestinal digestion of bovine beta-casein variants and milk-based infant formulas. Food Chem 110(4):897–903

De Noni I, Cattaneo S (2010) Occurrence of b-casomorphins 5 and 7 in commercial dairy products and in their digests following in vitro simulated gastro-intestinal digestion. Food Chem 119:560–566

Elliott RB, Harris DP, Hill JP et al (1999) Type (insulin dependent) diabetes mellitus and cow milk: casein variant consumption. Diabetologia 42(3):292–296

Elliott RB, Martin JM (1984) Dietary protein: a trigger of insulin-dependent diabetes in the BB rat? Diabetologia 26(4):297–299

Elliott RB, Wasmuth H, Hill J et al (1996) Diabetes and cows' milk. Lancet 348:1657

Elliott RB, Wasmuth HE, Bibby NJ et al (1997) The role of beta-casein in the induction of insulin-dependent diabetes in the non-obese diabetic mouse and humans. In: Seminar on milk protein polymorphism, IDF Special Issue 9702. International Dairy Federation, Brussels, pp 445–453

European Food Safety Authority (EFSA) (2009) Review of the potential health impact of β-casomorphins and related peptides. Sci Rep 231:1–107

Ganguly I, Kumar S, Gaur GK et al (2013) Beta-casein (CSN2) polymorphism in Ongole (Indian zebu) and Frieswal (HF × Sahiwal crossbred) cattle. Indian J Biotechnol 12(2):195–198

Haileselassie SS, Lee BH, Gibbs BF (1999) Purification and identification of potentially bioactive peptides from enzyme-modified cheese. J Dairy Sci 82(8):1612–1617

Hamel U, Kielwein G, Teschemacher H (1985) Beta-casomorphin immunoreactive materials in cows' milk incubated with various bacterial species. J Dairy Res 52(1):139–148

Hernández-Ledesma B, Amigo L, Ramos M et al (2004) Release of angiotensin converting enzyme-inhibitory peptides by simulated gastrointestinal digestion of infant formulas. Int Dairy J 14(10):889–898

Hernández-Ledesma B, Quirós A, Amigo L et al (2007) Identification of bioactive peptides after digestion of human milk and infant formula with pepsin and pancreatin. Int Dairy J 17(1):42–49

Jarmolowska B, Sidor K, Iwan M et al (2007) Changes of β-casomorphin content in human milk during lactation. Peptides 28(10):1982–1986

Jinsmaa Y, Yoshikawa M (1999) Enzymatic release of neocasomorphin and β-casomorphin from bovine β-casein. Peptides 20(8):957–962

Kahala M, Pahkala E, Pihlanto-Leppaelae A (1993) Peptides in fermented Finnish milk products. Agr Food Sci Fin 2(5):379–386

Kim YS, Whistle BW, Kim YW (1972) Peptide hydrolases in the brush border and soluble fractions of small intestinal mucosa of rat and man. J Clin Invest 51:1419–1430

Koch G, Wiedemann K, Teschemacher H (1985) Opioid activities of human β-casomorphins. Naunyn Schmiedeberg's Arch Pharmacol 331(4):351–354

Kumar S, Singh RV, Chauhan A (2019) Molecular characterization of A1/A2 β-casein alleles in Vrindavani crossbred and Sahiwal cattle. Indian J Anim Res 53:151–155

Laugesen M, Elliott R (2003) Ischaemic heart disease, type 1 diabetes, and cow milk A1 β-casein. N Z Med J 116(1168):U295

Law J, Handrickman A (1997) Proteolytic enzymes of lactic acid bacteria. Int Dairy J 7(1):1–11

Lynch J (2015) A2 weighs up building baby formula plant to cash in on China demand. Sydney Morning Herald

Matar C, Goulet J (1996) β-Casomorphin 4 from milk fermented by a mutant of lactobacillus helveticus. Int Dairy J 6(4):383–397

McLachlan CNS (2001) β-Casein A1, ischaemic heart disease mortality, and other illnesses. Med Hypotheses 56(2):262–272

Monetini L, Cavallo MG, Manfrini S (2002) Antibodies to bovine β-casein in diabetes and other autoimmune diseases. Horm Metab Res 34(8):455–459

Padberg S, Schumm-Draeger PM, Petzoldt R et al (1999) The significance of A1 and A2 antibodies against beta-casein in type-1 diabetes mellitus. Deut Med Wochenschr 124(50):1518–1521

Raies MH, Kapila R, Kapila S (2015) Release of β-casomorphin-7/5 during simulated gastrointestinal digestion of milk β-casein variants from Indian crossbred cattle (Karan fries). Food Chem 1(168):70–79

Raies MH, Kapila R, Saliganti V (2014a) Consumption of β-casomorphins-7/5 induce inflammatory immune response in mice gut through th2 pathway. J Funct Foods 8:150–160

Raies MH, Kapila R, Sharma R et al (2014b) Comparative evaluation of cow β-casein variants (a1/a2) consumption on th2-mediated inflammatory response in mouse gut. Eur J Nutr 53(4):1039–1049

Raies MH, Kapila R, Shandilya U (2012) Detection of A1 and A2 genetic variants of β-casein in Indian crossbred cattle by PCR-ACRS. Milchwissenschaft 67(4):396–398

Ramesha KP, Akhila R, Basavaraju M et al (2016) Genetic variants of β-casein in cattle and buffalo breeding bulls in Karnataka state of India. Indian J Biotechnol 15(2):178–181

Rando A, Di Gregorio P, Ramunno L et al (1998) Characterization of the CSN1AG allele of the bovine alpha s1-casein locus by the insertion of a relict of a long interspersed element. J Dairy Sci 81:1735–1742

Reichelt KL, Knivsberg AM (2003) Can the pathophysiology of autism be explained by the nature of the discovered urine peptides. Nutr Neurosci 6(1):19–28

Rokka T, Syväoja EL, Tuominen J et al (1997) Release of bioactive peptides by enzymatic proteolysis of lactobacillus GG fermented UHT milk. Milchwissenschaft 52:675–678

Schieber A, Brückner H (2000) Characterization of oligo-and polypeptides isolated from yoghurt. Eur Food Res Technol 210(5):310–313

Sun Z, Cade R (2003) Findings in normal rats following administration of gliadorphin- 7(GD-7). Peptides 24(2):321–323

Sun Z, Zhang Z, Wang X et al (2003) Relation of β-casomorphin to apnea in sudden infant death syndrome. Peptides 24(6):937–943

Svedberg J, De Haas J, Leimenstoll G et al (1985) Demonstration of beta-casomorphin immuno-reactive material in in-vitro digest of bovine milk and in small intestine contents after bovine milk ingestion in adult human. Peptides 6:825–830

Teresa, O (2011) "A2 Deal has Milk pouring into UK". The Australian. p 41

Thorsdottir I, Birgisdottir BE, Johannsdottir IM et al (2002) Different β-casein fractions in Icelandic versus Scandinavian cow's milk may influence diabetogenicity of cow's milk in infancy and explain low incidence of insulin-dependent diabetes mellitus in Iceland. Pediatrics 106(4):719–724

Trompette A, Claustre J, Caillon F (2003) Milk bioactive peptides and beta-casomorphins induce mucus release in rat jejunum. J Nutr 133(11):3499–3503

Yin H, Miao J, Ma C et al (2012) β-Casomorphin-7 cause decreasing in oxidative stress and inhibiting NF-κB-iNOS-NO signal pathway in pancreas of diabetes rats. J Food Sci 77(2):C278–C282

Yin H, Miao J, Zhang Y (2010) Protective effect of β-casomorphin-7 on type 1 diabetes rats induced with streptozotocin. Peptides 31(9):1725–1729

Zoghbi S, Trompette A, Claustre J et al (2006) Beta-casomorphin-7 regulates the secretion and expression of gastrointestinal mucins through a mu-opioid pathway. Am J Physiol Gastrointest Liver Physiol 290(6):G1105–G1113

Chapter 3
β-Casomorphin I

Abstract β-Casomorphins (BCMs) are the opioid peptides that bind μ-receptors and vary in number and sequence of amino acids. BCM-7 is the most abundant with prolonged effect while BCM-5 is the most active among BCM peptides. These peptdies have tyrosine at the N-terminal and another aromatic amino acid residue (Phe or Tyr) at third or fourth position. BCMs has been detected in raw cow milk (BCM-7), pasteurized milk (BCM-4) and fermented milks (BCM containing peptide sequences). The presence of BCM-9, BCM-7 and BCM-3 has been depicted from cheese samples (Brie, Cheddar, Probiotic Edam and Parmigianino Reggiano). BCM-5 and BCM-7 were detected in hydrolysed infant formulas (Humana, commercial and experimental milk-based infant formulas). B variant of β-casein depicted highest BCM-7 release followed by A1 variant on SGID. We also checked in our laboratory the release of BCM-7 from homozygous A1 and heterozygous variants of β-casein. However, the releases of these peptides in vivo from human gut or blood haven't been observed.

3.1 Structure of Bovine β-Casomorphins

Morphine is an alkaloid, however, β-casomorphins (BCMs) are peptides derived from β-casein protein of cow milk. Both the compounds have tendency to bind the same type of receptors i.e., mu (μ) receptors to show numerous biological effects. The active ligand molecules bind to their receptors based on the structural similarity. The particular ligand binds its cognate receptors when there is a complementary structure or spatial arrangement of molecules between two to form non-covalent interactions. For two ligand molecules like morphine and BCM to bind the same receptors, the two molecules must not only resemble in the steric, but should have similar electronic properties. In other words, the two compounds must have similar conformations that are thermodynamically stable and have ability to manifest same biological interaction with the same receptor that may be structurally unknown (Brandt et al. 1994). This structure function correlation can be explained by comparing the structure and activities (opioid) of BCM and morphine. For this instance, the opioid activity refers to the capacity of these compounds to interact and bind

© Springer Nature Singapore Pte Ltd. 2020

M. R. Ul Haq, *β-Casomorphins*, https://doi.org/10.1007/978-981-15-3457-7_3

with the opioid receptors. The opioid receptors are divided into subclasses as mu (µ1), mu (µ2) and delta (δ). Morphine has the highest specificity for mu (µ2), nevertheless, it may bind mu (µ1) also. There are different types of BCMs that depend on their amino acid sequence or no. of amino acids. The most active among bovine BCMs is the BCM-5. They do have affinity for mu (µ) receptors but this is lesser than morphine, and for it is lesser for delta (δ) receptors compared to former. The affinity between these peptides and kappa (κ) receptors seem to be negligible (Liebmann et al. 1991). A similar compound known as morphiceptin shows higher specificity towards µ-receptors, however, it demonstrates lower degree of conformational freedoms compared to BCM-5.

Molecular graphics (MEPS) is a branch of biophysical chemistry that establishes the correlation between the structure of a compound and related opioid receptor properties. This has been applied for the compounds like morphine, BCMs and morphiceptin to study their opioid action. This is a graphical representation that shows the electrostatic interactions between a unit charge (positive, proton) and the molecules under analysis. From this graphs certain conclusions are drawn including the electrostatic interactions between certain regions of a compound and the nucleophilic or electrophilic regions of the receptor. For example, whenever there is an electrostatic potential with negative sign, it implies a region of attraction for two positively charged groups (Brantl et al. 1979). The observations taken by these researchers from this technique explained that the position of the nitrogen in the opioid compound morphine in the ring (aromatic) is one of the key elements that are significant for manifestation of opioid activity. The existence of the hydroxyl group (phenolic) enhances the receptor affinity; nevertheless, it is not essential for expression of this activity (opioid activity). Reports have also clearly demonstrated that the other ring (benzene ring or F ring) is of considerable significance (Brandt et al. 1994). It is noteworthy that the aforementioned elements discussed are present in the BCMs, therefore, these peptides have enough chances to bind mu (µ) receptors and demonstrate opioid activity. The other factor that determines the opioid activity or binding of the opioid peptide with the opioid receptor is the hydrophilic and hydrophobic interactions.

From this perspective, Bakalkin and coworkers in 1992 analyzed the opioid peptides and found that these compounds have both hydrophobic and hydrophilic regions and in this regard grouped these amino acids in four classes as following (Bakalkin et al. 1992).

- **Group P:** *hydrophobic amino acid residues Pro, Tyr, Ile, Val, Trp and Cys*
- **Group G:** *hydrophilic amino acid residues Gly and Asn*
- **Group F:** *Phe, Leu, Ala, Met*
- **Group B:** *Asp, Glu, His, Gln*

The P group precedes G group that in turn precedes group F. The latter precedes Arg, Thr, Ser, and Lys (hydrophilic amino acid residues in R group) that in turn precedes P group. The researchers finally concluded that there existed a correlation between the amino acid sequence in the peptide (primary structure) and the activity (biological) in the family of opioid peptides. It is basically the properties of these amino acid sequences that determine the contents of few fragment pairs besides the compactness of their organization in a peptide (Bakalkin et al. 1992).

These observations apply directly to the replicated assessment of the hydrophobic index of the peptides derived from milk. Most of these opioid peptides have a hydrophobicity of about 2 kcal/mol and an aromatic hydrophobicity of about 0.6 kcal/mol. Overall, a correlation establishes quite well with the molecular weight of these opioid peptide. An inverse relation develops between molecular weight and hydrophobicity, i.e., the peptides with highest molecular weight is inversely proportional to the lowest hydrophobicity. This is in agreement with the data of Bakalkin (1992) that proves the existence of hydrophobic segments in the exogenous (exorphins) and endogenous (endorphins) (Bakalkin et al. 1992).

In a similar manner, a correlation between the opioid receptor affinity and aromatic hydrophobicity is established. For instance, to study the ability of opioid peptides to bind with these receptors, Bakalkin and coworkers in 1992 used pharmacological studies in guinea pig ileum and expressed the values in terms of IC50. They also performed binding assays (biochemical) and expressed the results in Ki. The IC50 values expressed as –log [IC50] (M) were calculated as 6.2 (morphiceptin), 5.6 (BCM-5) and 4.2 (α-casein) that correlates well with the aromatic hydrophobic index of these compounds 1.380 kcal/mol, 1.104 kcal/mol and 1.380 kcal/mol, respectively. In conclusion, the efficient binding and activity of an opioid peptide with its receptors requires the donor-acceptor arrangement in addition to a hydrophobic site. The side terminal residues of these peptides besides C terminal may work in the donor-acceptor arrangement. However, the presence of a hydro-philic/phobic equilibrium in opioid peptides could additionally support this conjugation.

There are many BCMs, however, the most prevalent or active natural BCMs are BCM-5, 7 and 9. The most abundant BCM is BCM-7 and most active is BCM-5. Theoretically it is possible that BCM-5 may be produced from BCM-7 by the simulated gastrointestinal digestion (SGID) in vitro or during digestion in humans. Wasilewska and coworker in 2011 detected the presence of BCM-5 in the serum taken from babies whose mothers take cow milk (Wasilewska et al. 2011). However, the formation of BCM-5 in either mothers or babies couldn't be established. The BCM that comes on the second position from the potency perspective is BCM-7, the latter has been correlated with incidence of various human complications. This peptide has been detected in human jejunal, blood of human infants, and urine of children (Wasilewska et al. 2011; Sokolov et al. 2014).

3.2 Classification of Bovine β-Casomorphins

BCMs are the opioid peptides released from the β-casein of milk protein. They are classified on the number of amino acid residues that ranges from 4 to 21 as depicted in Table 3.1. The common feature is that they all start with tyrosine from the N-terminal and occupies position 60 (Kostyra et al. 2004). These peptides are

Table 3.1 Different types of bovine BCMs

BCMs	Position in Bovine β-casein	Sequence
BCM-4	60–63	Tyr-pro-Phe-pro
BCM-5	60–64	Tyr-pro-Phe-pro-Gly
BCM-6	60–65	Tyr-pro-Phe-pro-Gly-[pro]
BCM-7	60–66	Tyr-pro-Phe-pro-Gly-[pro]-Ile
BCM-8	60–67	Tyr-pro-Phe-pro-Gly-pro-Ile-[pro]
BCM-9	60–68	Tyr-pro-Phe-pro-Gly-pro-Ile-[his]-Asn
BCM-13	60–72	Tyr-pro-Phe-pro-Gly-pro-Ile-[his]-Asn-Ser-Leu-pro-Gln
BCM-21	60–81	Tyr-pro-Phe-pro-Gly-pro-Ile-[his]-Asn-Ser-Leu-pro-Gln-Asn-Ile-pro-pro-Leu-Thr-Gln-Thr

derived from exogenous food elements like cow milk and show morphine-like actions, therefore, these peptides are also known as exorphins or formons i.e., food derived hormones, as they demonstrate hormone like actions. The common characteristic feature among these exorphins is presence of tyrosine amino acid residues at the N-terminal and another aromatic amino acid residue (Phe or Tyr) at third or fourth position. These peptides are regarded as very stable, this is due to the fact that these peptides are rich in proline residues, therefore resistant to enzymatic hydrolysis. Nevertheless, these are substrates for dipeptidyl peptidase IV (DPP IV). The latter is secreted as brush border enzyme and classified in prolyl oligopeptidase (PO) family.

3.3 Production of β-Casomorphins

3.3.1 Cow Milk (Whole)

Cieslinska and coworkers in 2007 reported the detection of BCM-7 in whole or raw cow milk (Cieslinska et al. 2007). This finding however raised many quires, like detection method employed and health of the dairy cows etc. They used only HPLC/ UV for analysis of this peptide that isn't sufficient. The former method should be followed by ELISA and mass spectroscopy for confirmation. It is quite possible that the peptides have same hydropathy index often eluting with co-eluting peaks. Moreover, there isn't a single study available that might have confirmed the results for presence of BCM-7 in raw or unprocessed cow milk. Furthermore, the health of the dairy cows from which the unprocessed milk was procured was not mentioned in the study. Quite evidently, the presence of somatic cells in the milk is a normal process. This is called somatic cell count (SSC). This count however increases in the diseased conditions like sub-clinical and clinical mastitis. This increase in SSC leads to increased expression of proteolytic enzymes in the milk that in turn leads to hydrolysis of proteins. Therefore there is a probability that in these conditions there is production of peptides like BCMs or the precursors from which these can be obtained later.

Although model systems have been employed for determination of activity and specificity of elastase and cathepsins, however, their exact levels of expression in milk hasn't been studied. The cathepsin B activity is present in the somatic cells (O'Driscoll et al. 1999). In vitro studies of digestion of β-casein have reported that the cathepsin B activity is similar to that of plasmin and cell envelope proteinases (CEP). These enzymes are expressed by various *Lactococcus lactis* strains. Additionally cathepsin B has been found to release BCM-10 from the β-casein of cow milk (Considine et al. 2004).

In an another trial by Wedholm and coworkers in (2008) found a good correlation between elastase and cathepsins activity and hydrolysis of β-casein of cow milk, however, the study couldn't find any release of BCM. This observations were taken in sub-clinical mastitis with *Streptococcus uberis* of dairy cows having SSC > 500,000 cells/mL (Wedholm et al. 2008).

In another study, the activity of various proteinases was analyzed at acid and physiological pH for 24–216 h at an incubation temperature of 37 °C. The milk taken for study was unprocessed and produced from every quarter of the mammary gland of a mastitis and healthy cows with a SCC of $0.4–11.6 \times 10^6$ and $0.6–1.3 \times 10^3$ respectively. The merit of this study was the validated system used for detection of peptides MALDI-MS and MS/MS. The results indicated presence of peptides that have hydrolytic site located towards the C-terminus but not in the BCM region (Napoli et al. 2007).

The presence of somatic cells in milk and their proteinase activity may be involved in the hydrolysis of β-casein, and may also occur in the ultra-heated (UHT) milk. However, there is a possibility that these enzymes may be inactivated by this heat application. Nevertheless, to the best of our knowledge, there isn't any study available that may have demonstrated the heat inactivation of enzymes in UHT-milk. Consequently Gaucher and coworkers in 2008 found that β-casein is hydrolyzed even in UHT milk, (heated at 140 °C for 4 s and the stored at 20 °C for 6 months). They used advanced instrumentation (LC-MS/MS) for analysis. Although the investigators couldn't find any BCM, nevertheless, the various BCM precursors were detected (F4–68, F54–69, F55–65, F55–68, F57–68) with the enzymatic activity of elastase, cathepsin B and cathepsin G on β-casein (Gaucher et al. 2008).

3.3.2 *Fermented Milk*

The production of BCMs in general and BCM-7 in particular has been studied in a number of fermented milk products like fermented milk, cheese and yoghurt. The actual data relating the release and production of these peptides is scanty. The lack of information in this perspective is due to the complex fermented system and detection in such a complex system, the number of peptides produced from such diverse protein family of cow milk. Nevertheless, there have been some trials in this regard where purified caseins, β-casein or commercially synthesized BCMs were hydrolyzed with enzyme systems derived from bacterial strains for analysis of BCMs, their precursors or other peptides. It was Hamel and co-researchers for the first time in 1985 who detected BCM immunoreactive substances from milk incubated with lactic acid bacteria (LAB). Already it is well-established that these bacteria secrete CEP and other peptidases (intracellular). CEP produces the long peptides from the milk proteins; however, intracellular peptidases further digest these oligopeptides into smaller fragments in addition to free amino acids (Hamel et al. 1985).

However, almost more than one decade ago in 2007, researchers isolated LAB (21 strains) from fermented milk, performed in vitro trials with these bacteria with sodium caseinate solution. They reported that β-casein (80–90%) is digested with these bacteria when incubated for 72–96 h (Tzvetkova et al. 2007). The findings were in agreement with Mierau and coworkers found that proteinases derived from LAB (*Lactococcus lactis* subsp. *Cremoris*) digested almost 40% of β-casein and produced more than 100 oligopeptides (Mierau et al. 1997).

The production of BCMs from systems utilizing LAB is improbable, because these bacteria secrete dipeptidyl peptidase like X-prolyl dipeptidyl aminopeptidase (PepX) (Gobbetti et al. 2002), the latter has high specificity for proline residues. The BCMs being rich in this amino acid has high probability of being digested to dipeptides like X-Pro from the N-terminal. The X-Pro is essential for the expression of the biological activities of BCM opioid peptides and hence this hydrolysis with pepX make them biological inactive. The wild type *Lactobacillus helveticus* L89 either in wild form or mutant stains (pepX deficient) were found to breakdown BCM-7 to BCM-4 fully or partially at 37 °C incubated for 120 min. The more advanced techniques like LC/MS further demonstrated the production of BCM-4 but not BCM-7 in pasteurized milk (65 °C for 30 min). The milk was fermented with *Lactobacillus helveticus* L89 deficient in PepX (Matar and Goulet 1996). Further studies showed the inhibition of peptidases (PepO and PepN) but not PepX from *Lactococcus lactis* ssp. *lactis* MG1363. It was also observed that BCM-7 is cleaved by pepX (Stepaniak et al. 1995). BCM-7 as compared to BCM-3 and BCM-5 was found to be more resistant to proteolysis after 6–15 weeks of incubation with *Lactococcus Cremoris* (Muehlenkamp and Warthesen 1996). Different types of peptides are produced by different starter cultures in fermented milk. It is well-established that *Lb. delbrueckii* subsp. *bulgaricus* and *S. thermophilus,* when used together in the preparation of yoghurt produce peptides which encourage the growth of the assorted culture. When fermentation is carried out for short time, the breakdown of the LAB is improbable and which in turn is unlikely that peptidase activity (intracellular) to remarkably contribute to hydrolysis during preparation of yoghurt (Meisel and Bockelmann 1999). The capacity of the bacteria to release peptides from protein (proteolysis) during fermentation of milk with *L. delbruekii* ssp. *bulgaricus* Lb1466 and *S. thermophilus* St1342 is highest during the first 24 h and this decreases subsequently during storage (28 day, 4 °C) (Donkor et al. 2007).

Schieber and Bruckner in 2000 reported the presence of BCM containing peptide sequences from 57–68 and 57–72 from A1 variants of β-casein with *L. delbruekii* ssp. *bulgaricus* Lb1466 and *S. thermophilus* St1342. These bacterial strains were used for fermentation of skim milk to yoghurt, the milk was first heated to 90 °C and then fermented at 44 °C for 3 h. the analysis was carried out with HPLC followed by mass spectroscopy (Schieber and Brückner 2000). However, when milk (Finnish) was fermented to yoghurt with mixed culture of *L. bulgaricus* and *S. thermophilus,* results indicated the production of peptides from N and C end of β-casein, on contrary to the above observations, no BCMS were detected (Kahala et al. 1993). Donkor and coworkers in 2007 observed a higher proteolytic activity in fermented milk comprising probiotic bacteria (*B. lactis* B94, *L. acidophilus* L10 and *L. casei* L26) and yoghurt starter culture. However this proteolytic activity was found very low when fermentation included yoghurt culture only (Donkor et al. 2007). The probiotic bacteria (*Bifidobacterium* spp. and *L. acidophilus)* were found to have low proteolytic efficiency compared to yogurt bacteria (*L. delbrueckii* ssp. *Bulgaricus* and *S. thermophilus*) (Shihata and Shah 2000). The authors observed that preparation of fermented milk with probiotic bacteria affects the peptide profile. Consequently the peptide profile may affect the production of BCMs during SGID. When fermentation of milk (UHT milk, 142 °C for 3–4 s) was carried out with the probiotic *Lactobacillus* GG strain and followed by pepsin and trypsin digestion, results indicated the release of protein fragments similar to peptide sequences BCM-11 and BCM-4. Nevertheless, these peptides were not detected in fermented milk without subsequent enzyme hydrolysis with pepsin and trypsin digestion (Rokka et al. 1997).

3.3.3 Cheese

The cheese is considered among the complex food system, therefore, difficult from peptide analysis perspective. Nevertheless, there are many reports available that have shown the production of BCMs from such a complex system. The release of peptides, extent and pattern of cheese proteolysis depends on the following parameters.

- Type of treated milk (pasteurized or unprocessed)
- Coagulation type (animal or vegetable coagulant used),
- Curd handling (cooking, salting, pH at draining),
- Starter culture
- Ripening conditions (pH, time, temperature, humidity, secondary microflora).

The extent of hydrolysis is highest in the ripening stage, the caseins are cleaved by peptidase, coagulant and other proteinases (starter or non-starter). As aforementioned, BCMs in general and BCM-7 in particular and some related peptides inhibit the action of endopeptidase as well as exopeptidases secreted by LAB (Stepaniak et al. 1995). These factors may therefore affect the production and degradation of

BCM-7 that may therefore be estimated during real cheese preparation circumstances. In this perspective, it is unlikely for chymosin to release BCM independent on the pH and salt concentration values. Moreover, plasmin proteolysis is unable to release bioactive peptides (McSweeney 2004). As discussed earlier, the peptidases from starter and non-starter LAB have the tendency to cleave the proline containing peptides like BCMs, the latter are formed by the milk (cheese or curd) containing proteolytic systems. During cheese ripening, there is degradation of LAB, the released peptidase therefore contribute to proteolysis at this stage (Meisel and Bockelmann 1999). This ripening is assisted by additional proteases released from fungus that are used in the preparation of certain types of cheese. Although the mechanism of the proteinases for the cheese is well-established, still it is very unlikely to predict production of BCM-7 from cheese. There are very few reports available that show release of this peptide from cheeses. Taking into account the analysis of amino acid composition, Jarmolowska and coworkers in 1999 reported the production of BCM-7 from Brie cheese samples and the content obtained was from 5 to 15 mg/kg cheese (Jarmolowska et al. 1999). Earlier, Muehlenkamp and coworkers in 1996 used HPLC-UV for peptide analysis and couldn't find any release of BCMs from the cheese samples used by the aforementioned studies. While using other cheese samples like Cheddar, Blue, Limburger and Swiss type (60 day ripened), still the investigators were unable to report presence of BCM-7 (Muehlenkamp and Warthesen 1996). In addition, BCM-7 was not detected in other cheese varieties having different extents of proteolysis: Cheddar, Swiss type (60 day ripened), Blue and Limburger. The authors reported that this peptide was resistant to enzyme hydrolysis secreted by *Lactococcus Cremoris* at pH 5.0–5.2 and high salt concentration (sodium chloride, 5%). When the pH was kept constant and the salt concentration was decreased to 1.5%, this peptide degraded (50%) or completely after 15 weeks. Therefore, keeping the pH and salt concentration under consideration, it is reasonable to put forward that this peptide (BCM-7) is likely to be broken down to some extent in cheese (Cheddar) by the starter culture (Muehlenkamp and Warthesen 1996). The peptides with different activities including BCM-7 were detected in enzyme-modified Cheddar cheese (EMC) that was manufactured with *Bacillus subtilis* extracted neutral protease (Neutrase®). However, the accurate amounts of BCM-7 produced in this cheese were not observed. The researcher used more reliable technique like LC-MS/MS for analysis of peptides. However, the same investigators couldn't find BCM-7 in the same type of cheese when prepared with crude extract of *L. casei*. These observation therefore support the earlier findings where the proteolytic action of dipeptidyl peptidase (proline specific peptidases) secreted by LAB has been demonstrated (Haileselassie et al. 1999). BCM-9 (60–68 from β-casein) has been detected in Cheddar cheese, the latter secreted by LAB in the same cheese is found to have little or no pepX activity (Singh et al. 1997). There are other reports available that have not detected presence of BCM-7 in these cheeses but the long peptide precursor sequences incorporated in it the BCM-7 sequences. Accordingly, Addeo and coworkers in 1992 demonstrated the presence of peptide precursors (20–21 amino acid residues) possessing in it BCM-7 sequence in ripened Parmigianino Reggiano cheese. The investigators are of the view that the existence of this long peptide establishes the resistance of cow β-casein for further hydrolysis. Further, it is reported that casein hydrolysis occurs to such an extent that 20% of nitrogen is in the form of free amino acids (Addeo et al. 1992). Stepaniak and its fellow workers in 1995 purified peptides (58–72) from A1 variant of β-casein in the extract that is water soluble from Cheddar and Jarlsberg cheeses (Stepaniak et al. 1995). Other workers also demonstrated the release the aforementioned peptide (58–72) at the Crescenza ripening. This peptide was found to be resistant to further hydrolysis and was not cleaved with PepO and PepN activities (Smacchi and Gobbetti 1998). As noted earlier, the β-casein (58–72) is observed to inhibit several protein degrading enzymes including PepN (general aminopeptidase), PepO (endopeptidase) and PepX from *Lc. Lactis* ssp. *lactis* MG1363 (Stepaniak et al., 1995). BCM-9 was also detected in Gouda (Saito et al. 2000). It was Toelstede and coworkers in 2008 who reported presence of BCM-9 and BCM-10 in cheese using the advanced detection method (LCMS-TOF and LC-MS/MS). They reported release of BCM-9 from both the variants of β-casein (A1/A2), however, BCM-10 was found to be released from only A1 variants of β-casein (Toelstede and Hofmann 2008). Gagnaire and co-investigators in 2001 didn't find any release of either BCMs or their precursors in Emmental cheese. The authors speculated that high temperature (curd cooking), CEP, cathepsin and plasmin resulted in the inactivation of coagulant that was actually responsible for hydrolysis of β-casein (Gagnaire et al. 2001). The presence of BCM-3 (but not BCM-5) was detected in Edam and probiotic Edam cheeses manufactured from Bifidobacterium with HPLC/UV detection system (Sabikhi and Mathur 2001). When the water-soluble fraction of Caprino del Piemonte cheese was analyzed with LC-MS, the authors reported release of a peptide sequence (YPFTGPIPN) ranging between 60 and 80 from β-casein (Rizzello et al. 2005).

3.3.4   Infant Formulas

The literature available that shows the presence of BCMs in infant formulas is scanty. Jarmolowska in 2007 checked the infant formula "Humana" for the presence of opioid peptides. They found that four peptides were released in the extract (pepsin) and pepsin-trypsin hydroxylate. The peptides detected were BCM-5, casoxin C and 6 and lactoferroxin A. The activity (opioid) of these extracts was checked on the motor activity on intestine (rabbit). The amount of BCM-5 was found in the peptide and pepsin-trypsin hydroxylate as 0.39, 43.11 μg/mL, casoxin 6 as 0.22, 6.21 μg/mL, casoxin C 0.075, 23.33 μg/mL and lactoferroxin A 0.16, 11.40 μg/mL. The authors found that the opioid peptides were found at a higher concentration in pepsin hydrolysate compared to non-pepsin hydrolysate in "Humana" infant formula (Jarmolowska et al. 2007). In 2008, the production of BCM-5 and BCM-7 was analyzed during SGID in commercial milk-based infant formulas ($n = 6$) and experimental infant formulas ($n = 3$). The detection of peptides was done by HPLC-UV followed by LC/MS. The SGID was carried out with pepsin (pH 2.0, 3.0 and 4.0) and the further digestion by Corolase PPTM. Quite interestingly, the β-casein variants A1, A2, and B were existing in all the commercial infant formulae (IFs) that were subjected to SGID. Consequently, the production of BCM-7 (0.02–0.37 nmol) were obtained per milliliter of infant formula (De Noni 2008). More recently, BCMs (8–11) have been detected in human milk before β-casein gets digested. It is suggested that the proteases present in milk lead to the proteolysis of β-casein to BCMs (Enjapoori et al. 2019).

3.4 BCM Release During SGID

The SGID employs two main characteristic features including the physiologically important enzymes that are significant from metabolic perspective. The other system is the physiological conditions used to main the environment like pH and temperature. The former is maintained separately for the enzymes knowing their optimum pH and the latter is maintained around 37 °C. The time of hydrolysis used in this method is correlated well with the physiological digestion processes. The release of BCMs from either extracted caseins or more specified β-casein has been tried in many studies. The pepsin is generally used to cleave the protein into smaller fragments while the trypsin and chymotrypsin further aid in this process by cleaving the peptides. The use of other hydrolytic enzymes like amino peptidases and carboxypeptidases leads to observations of more reproducible and reliable results. The significance of pepsin in SGID of β-casein was well demonstrated by Jinsmaa and coworkers in 1999 where production of BCM-9, BCM-13 and BCM-21 or (Val59)-BCMs occurred with pepsin hydrolysis of the Leu58-Val59 linkage in β-casein. Further production of BCM-7 (17 mmol/mol β-casein) in little amounts were observed with elastase-leucine aminopeptidase hydrolysis without the preliminary pepsin hydrolysis. Moreover, there wasn't any production of BCM-7 when either chymotrypsin or trypsin was employed instead of pancreatin and further pepsin (individually) was also unable to release BCM-7. In 1999, Macaud and co-researchers detected release of BCM-3, however, no BCM-7 was detected on pepsin hydrolysis while employing spectra (second-order) and ultraviolet-visible comparisons (Macaud et al. 1999). Similarly, the other workers couldn't find any release of BCM-7 from either of the variants of β-casein (A1 or A2) after pepsin hydrolysis in SGID. They performed hydrolysis with pepsin for 10–60 min at pH 2.0. The data analysis was performed with more sophisticated instruments (LC-MS) (Schmelzer et al. 2007). In 2008, De Noni and co-workers, more specifically tried to evaluate the release of BCMs from different variants of β-casein including A1, A2 and B types. They performed SGID using pepsin and Corolase PPTM. The pepsin hydrolysis was carried out at different pH (2.0, 3.0 and 4.0). The different variants of β-casein were isolated from raw milk and dispersed in UF permeate (cow milk). Although the pH change had no effect, the B variant of β-casein depicted highest BCM-7 release (5–176 mmol/mol casein), this was followed by release from A1 genetic variant. The release of BCM-7 from A2 genetic variant couldn't be elucidated. Moreover, the release of BCM-5 couldn't be detected from either of the genetic varaints of β-casein or change in pH during SGID (De Noni 2008). The release of BCM-7 from both the A1 and A2 variants of β-casein was observed by Cieslinska et al. (2007) from unprocessed milk, however, the release from the former was almost four times higher compared to latter after digesting these with pepsin at pH 2.0 for 24 h. This finding however raised many quires, like detection method employed and health of the dairy cows etc. They used only HPLC/UV for analysis of this peptide that isn't sufficient. The former method should be followed by ELISA and mass spectroscopy for confirmation. It is quite possible that the peptides have same hydropathy index often eluting with co-eluting peaks. Moreover, there isn't a single study available that might have confirmed the results for presence of BCM-7 in raw or unprocessed cow milk. Furthermore, the health of the dairy cows from which the unprocessed milk was procured was not mentioned in the study (Cieslinska et al. 2007). The release of BCM-7 from the β-casein extracted from the buffalo milk was studied by Petrilli et al. (1984) using the enzymes pepsin, chymotrypsin, trypsin, elastase, carboxypeptidase A and B, and leucine aminopeptidase (LAP). They analyzed the data with HPLC. It was observed that the gastric and pancreatic enzyme combinations couldn't release BCM-7. However, the enzyme digests yielded the precursor of 10 amino acids from 59–68 with internal BCM-7 sequence. Moreover, the treatment of these digest with the brush border enzymes of rabbit intestine were unable to produce BCM-7 (Petrilli et al. 1984). The same investigator extended the research in 1987 and tried to incubate the BCM-7 sequence containing fragment from 49–68 of buffalo β-casein with pancreatic juice (pig). The analysis was carried out with fast atom bombardment MS, the authors couldn't find the release of either BCM or morphiceptin (Petrilli et al. 1987). The BCM-7 was recovered, however, the release of this peptide was not quantified in the infant formula (reconstituted and UF permeate) that was digested with pepsin (pH, 3.5) followed by hydrolysis with Corolase PPTM (Hernández-Ledesma et al. 2004). However, the same researcher couldn't detect the same peptide in pepsin digested (pH, 3.5) infant formula and pancreatin (porcine) (Hernández-Ledesma et al. 2007). Nevertheless, the peptide BCM-9 was found in the same infant formula by UV-HPLC. Jarmolowska and coworkers in 2007 observed the release of BCM-5 in the infant formula digested by pepsin and pepsin-trypsin. The yield of BCM-5 in IF extracts (0.39 µg/mL) and its hydrolysate (43.11 µg/mL) was also observed. However, there wasn't any production of BCM-7 in the extracts before or after the SGID (Jarmolowska et al. 2007). During my PhD programme at National Dairy Research Institute (NDRI), we also checked the release of BCM-7 and BCM-5 from A1A1, A1A2 and A2A2 genetic variants of β-casein. This release was assessed after SGID using pepsin, trypsin, chymotrypsin, elastase and leucine amino peptidase. The three variants of β-casein were isolated from milk of crossbred cattle Karan Fries. These cattle were first screened for these genotypes using polymerase chain reaction-Amplification Creation restriction Digestion (PCR-ACRS). The detection of these peptides was carried out with analytical HPLC and the fractions were separated using preparative HPLC. The peptides were quantified with competitive

ELISA and sequences of the peptides were elucidated with MS-MS spectrometry. Among many fractions collected, the fraction B showed the presence of a peptide sequence with 14 amino acids (VYPFPGPIHNSLPQ). This 14 amino acid peptide has in its sequence encrypted an internal BCM sequence. When this 14 amino acid sequence was further digested with elastase and leucine amino peptidase, it revealed the presence of BCM-7 that was quantified with competitive ELISA as shown in Fig. 2.2. The amount of BCM-7 released from A1A1 was found to be 0.20 ± 0.02 mg/g β-casein that was almost 3.2 times more than heterozygous A1A2 variant of β-casein. On the other hand, as expected, the release of BCM-7 wasn't observed from A2A2 β-casein variant. Moreover, the release of BCM-5 wasn't detected from any of the variants of β-casein (Raies et al. 2015).

3.5 Processing of Cow Milk and BCM Release

To the best of the literature surveyed, there isn't any study available that correlates different factors or agents employed in milk processing with the release of BCMs. The well-established fact is that application of intense temperature or heat during preparation of yoghurt between 90–95 °C for duration of 5–20 min may prompt thermal alterations of protein biomolecules that actually contribute to the much anticipated characteristics of the ultimate item for consumption (i.e. viscosity, lack of syneresis). The increase in temperature (heating) may lead to denaturation and hence unfolding that ultimately leads to a change in protein secondary and tertiary changes. Therefore, the major milk proteins of milk i.e., caseins are more resistant to denaturation (temperature) compared to whey proteins (globular). The β-casein in casein fraction of cow milk forms α-helices that are short and therefore demonstrate less well-developed tertiary structure. This is for this reason that that this protein conformation changes least and hence shows lesser denaturation. This ultimately pays very least chances for alterations in the production of BCMs on heat treatments.

In 2007, Schmelzer reported the release of peptides from the β-casein during SGID. The authors couldn't observe any special cleavage pattern for pepsin predicted on secondary structure. Therefore, they concluded that pepsin hydrolyzes the β-casein in a similar fashion for both native and heat treated β-casein forms of cow milk (Schmelzer et al. 2007). It has been observed that milk heated strongly increases the establishment of complexes of kappa (κ)-casein or β-lactoglobulin on the external surface of micelles (casein). The formation of this conjugates (casein micelle-κ/β-lactoglobulin) makes the casein molecules less accessible to the proteolytic enzymes and leads to lesser production of peptides (Enright et al. 1999). Moreover, the other investigators found that when the milk from cows and goats were incubated with duodenal and gastric juices, the heated milk was found to be more resistant to proteolysis compared to unprocessed milk (Almaas et al. 2006).

There are formations of covalent interactions like lactosylation and crosslinking between casein molecules or within the same molecules. These changes may lead to production of xenobiotics that in turn causes the β-casein molecules to aggregate or polymerize (Pellegrino et al. 1999). The other post-translational modifications like phosphorylation-dephosphorylation and deamidation of β-casein when milk is processed during yoghurt preparation (Van Boekel 1999). These modifications in β-casein with different processes and the susceptibility to cleavage by hydrolytic enzymes during fermentation are difficult to speculate. The effect of heat treatments on casein or β-casein of cow milk and release of BCMs has also not been established. To the best of our knowledge, there isn't report available that have correlated the effect of ultra-heat treatment with the release of BCMs. Moreover the sterilization (in-bottle) and the release of BCMs from β-casein of bovine milk protein haven't been established. The infant milk based formulas uses various ingredients that are produced industrially and produced with various heat treatments. Further these infant formulas are further subjected to heated treatments. In this regard, these heat treatments can damage the final product of infant formulas. In this regard, De Noni in 2008, studied the ultra-heat treatments at 156 °C for a short duration of 6–9 s and didn't find significant change in the release of BCMs with this treatment. The raw preparations and heat-treated infant preparations yielded comparable amounts of BCMs (De Noni 2008).

References

Addeo F, Chianese L, Salzano A et al (1992) Characterization of the 12% trichloroacetic acid-insoluble oligopeptides of Parmigiano-Reggiano cheese. J Dairy Res 59(3):401–411

Almaas H, Cases AL, Devold TG et al (2006) In vitro digestion of bovine and caprine milk by human gastric and duodenal enzymes. Int Dairy J 16(9):961–968

Bakalkin G, Ya Demuth HU, Nymberg F (1992) Relationship between primary structure and activity in exorphins and endogenous opioid peptides. FEBS Lett 310:13–16

Brandt W, Barth A, Höltje HD (1994) Investigations of structure-activity relationships of β-casomorphins and further opioids using methods of molecular graphics. In: Brantl V, Teschemacher H (eds) β-casomorphins and related peptides: Recent developments. VCH, Weinheim, pp 93–104

Brantl V, Teschemacher H, Hemschem A et al (1979) Novel opioid peptides derived from casein (β-casomorphins). Z Physiol Chem 360:1211–1216

Cieslinska A, Kaminski S, Kostyra E et al (2007) Beta-casomorphin 7 in raw and hydrolyzed milk derived from cows of alternative beta-casein genotypes. Milchwissenschaft 62(2):125

Considine T, Healy A, Kelly AL et al (2004) Hydrolysis of bovine caseins by cathepsin b, a cysteine proteinase indigenous to milk. Int Dairy J 14(2):117–124

De Noni I (2008) Release of beta-casomorphins 5 and 7 during simulated gastro-intestinal digestion of bovine beta-casein variants and milk-based infant formulas. Food Chem 110(4):897–903

Donkor ON, Henriksson A, Singh TK et al (2007) ACE inhibitory activity of probiotic yoghurt. Int Dairy J 17(11):1321–1331

Enjapoori AK, Kukuljan S, Dwyer KM et al (2019) In vivo endogenous proteolysis yielding beta-casein derived bioactive beta-casomorphin peptides in human breast milk for infant nutrition. Nutrition 57:259–267. https://doi.org/10.1016/j.nut.2018.05.011

Enright E, Patricia Bland A, Needs EC et al (1999) Proteolysis and physicochemical changes in milk on storage as affected by UHT treatment, plasmin activity and KIO3 addition. Int Dairy J 9(9):581–591

Gagnaire V, Molle D, Herrouin M (2001) Peptides identified during emmental cheese ripening: origin and proteolytic systems involved. J Agric Food Chem 49(9):4402–4413

Gaucher I, Mollé D, Gagnaire V et al (2008) Effects of storage temperature on physico-chemical characteristics of semi-skimmed UHT milk. Food Hydrocoll 22(1):130–143

Gobbetti M, Stepaniak L, De Angelis M et al (2002) Latent bioactive peptides in milk proteins: proteolytic activation and significance in dairy processing. Crit Rev Food Sci Nutr 42(3):223–239

Haileselassie SS, Lee BH, Gibbs BF (1999) Purification and identification of potentially bioactive peptides from enzyme-modified cheese. J Dairy Sci 82(8):1612–1617

Hamel U, Kielwein G, Teschemacher H (1985) Beta-casomorphin immunoreactive materials in cows' milk incubated with various bacterial species. J Dairy Res 52(1):139–148

Hernández-Ledesma B, Amigo L, Ramos M (2004) Release of angiotensin converting enzyme-inhibitory peptides by simulated gastrointestinal digestion of infant formulas. Int Dairy J 14(10):889–898

Hernández-Ledesma B, Quirós A, Amigo L et al (2007) Identification of bioactive peptides after digestion of human milk and infant formula with pepsin and pancreatin. Int Dairy J 17(1):42–49

Jarmolowska B, Kostyra E, Krawczuk S et al (1999) β-Casomorphin-7 isolated from brie cheese. J Sci Food Agric 79:1788–1792

Jarmolowska B, Szlapka-Sienkiewicz E, Kostyra E et al (2007) Opioid activity of humana formula for newborns. J Sci Food Agric 87(12):2247–2250

Kahala M, Pahkala E, Pihlanto-Leppaelae A (1993) Peptides in fermented Finnish milk products. Agr Food Sci Finland 2(5):379–386

Kostyra E, Sienkiewicz-Szapka E, Jarmolowska B et al (2004) Opioid peptides derived from milk proteins. Pol J Food Nutr 13(54):25–35

Liebbmann C, Schnitte M, Hartrodt B et al (1991) Structure-activity studies of novel casomorphin analogues: binding profiles μ1, μ2 and δ-opioid receptors. Pharmazie 46:345–352

Macaud C, Zhao Q, Ricart G (1999) Rapid detection of a casomorphin and new casomorphin-like peptide from a peptic casein hydrolysate by spectral comparison and second order derivative spectroscopy during HPLC analysis. J Liq Chromatogr Relat Technol 22(3):40–418

Matar C, Goulet J (1996) β-Casomorphin-4 from milk fermented by a mutant of lactobacillus helveticus. Int Dairy J 6(4):383–397

McSweeney PLH (2004) Biochemistry of cheese ripening. Int J Dairy Technol 57(2–3):127–144

Meisel H, Bockelmann W (1999) Bioactive peptides encrypted in milk proteins: proteolytic activation and thropho-functional properties. Antonie Van Leeuwenhoek 76(1):207–215

Mierau I, Kunji ERS, Venema G et al (1997) Casein and peptide degradation in lactic acid bacteria. Biotechnol Genet Eng Rev 14:279–301

Muehlenkamp MR, Warthesen JJ (1996) Beta-casomorphins: analysis in cheese and susceptibility to proteolytic enzymes from Lactococcus lactis ssp Cremoris. J Dairy Sci 79(1):20–26

Napoli A, Aiello D, Donna D et al (2007) Exploitation of endogenous protease activity in raw mastitic milk by MALDI-TOF/TOF. Anal Chem 79(15):5941–5948

O'Driscoll BM, Rattray FP, McSweeney PLH et al (1999) Protease activities in raw milk determined using a synthetic heptapeptide substrate. J Food Sci 64(4):606–611

Pellegrino L, Van Boekel M, Gruppen H et al (1999) Heat induced aggregation and covalent linkages in β-casein model systems. Int Dairy J 9(3–6):255–260

Petrilli P, Picone D, Caporale C et al (1984) Does casomorphin have a functional role? FEBS Lett 169(1):53–56

Petrilli P, Pucci P, Pelissier JP et al (1987) Digestion by pancreatic juice of a beta-casomorphin-containing fragment of buffalo beta-casein. Int J Pept Protein Res 29(4):504–508

Raies MH, Kapila R, Kapila S (2015) Release of β-casomorphin-7/5 during simulated gastrointestinal digestion of milk β-casein variants from Indian crossbred cattle (Karan fries). Food Chem 1(168):70–79

Rizzello CG, Losito I, Gobbetti M et al (2005) Antibacterial activities of peptides from the water-soluble extracts of Italian cheese varieties. J Dairy Sci 88(7):2348–2360

Rokka T, Syvaoja EL, Tuominen J et al (1997) Release of bioactive peptides by enzymatic proteolysis of lactobacillus GG fermented UHT milk. Milchwissenschaft 52:675–678

Sabikhi L, Mathur BN (2001) Qualitative and quantitative analysis of β-casomorphins in Edam cheese. Milchwissenschaft 56(4):198–200

Saito T, Nakamura T, Kitazawa H (2000) Isolation and structural analysis of antihypertensive peptides that exist naturally in gouda cheese. J Dairy Sci 83(7):1434–1440

Schieber A, Brückner H (2000) Characterization of oligo-and polypeptides isolated from yoghurt. Eur Food Res Technol 210(5):310–313

Schmelzer CEH, Schöps R, Reynell L (2007) Peptic digestion of β-casein time course and fate of possible bioactive peptides. J Chromatogr A 1166(1–2):108–115

Shihata A, Shah NP (2000) Proteolytic profiles of yogurt and probiotic bacteria. Int Dairy J 10(5–6):401–408

Singh TK, Fox PF, Healy A (1997) Isolation and identification of further peptides in the diafiltration retentate of the water-soluble fraction of cheddar cheese. J Dairy Res 64(03):433–443

Smacchi E, Gobbetti M (1998) Peptides from several Italian cheeses inhibitory to proteolytic enzymes of lactic acid bacteria, pseudomonas Fluorescens ATCC 948 and to the angiotensin i-converting enzyme. Enzym Microb Technol 22(8):687–694

Sokolov O, Kost N, Andreeva O et al (2014) Autistic children display elevated urine levels of bovine casomorphin-7 immunoreactivity. Peptides 56:68–71

Stepaniak L, Fox PF, Sorhaug T et al (1995) Effect of peptides from the sequence 58-72 of beta-casein on the activity of endopeptidase, aminopeptidase, and X prolyl-dipeptidyl aminopeptidase from Lactococcus Lactis ssp. Lactis MG1363. J Agric Food Chem 43(3):849–853

Toelstede S, Hofmann T (2008) Sensomics mapping and identification of the key bitter metabolites in gouda cheese. J Agric Food Chem 56(8):2795–2804

Tzvetkova I, Dalgalarrondo M, Danova S et al (2007) Hydrolysis of major dairy proteins by lactic acid Bacteria from Bulgarian yogurts. J Food Biochem 31(5):680–702

Van Boekel M (1999) Heat-induced deamidation, dephosphorylation and breakdown of caseinate. Int Dairy J 9(3–6):237–241

Wasilewska J, Sienkiewicz-Szlapka E, Kuzbida E (2011) The exogenous opioid peptides and DPPIV serum activity in infants with apnea expressed as apparent life threatening events (ALTE). Neuropeptides 45(3):189–195

Wedholm A, Moller HS, Lindmark-Mansson H (2008) Identification of peptides in milk as a result of proteolysis at different levels of somatic cell counts using LC MALDI MS/MS detection. J Dairy Res 75(1):76–83

Chapter 4
β-Casomorphin II

Abstract BCMs have demonstrated a role in the transport of amino acids and secretion of mucus through mucin production in the intestine. These peptides affect the postprandial metabolism, reduction in the fat intake, increase in the gastrointestinal transit time, electrolyte absorption modulation, inhibition in the gastric transit, up-regulation of DDPIV and myeloperoxidase activity in the intestine. Moreover, intake of A1 milk was related to elevated inflammatory response of gastrointestinal system, deterioration of PD3 symptoms, the delay in gastrointestinal transit and the reduced cognitive processing speed and accuracy. BCMs demonstrated a depression in respiratory frequency and tidal volume. BCMs showed sedative activity, analgesic and cardiovascular effects, production of hypotension and bradycardia. BCM-5 depicted ionotropic activity at lower and cardiodepressive response at higher doses. BCM-7 and BCM-10 showed a proliferation of lymphocytes at lower and stimulation at higher doses. BCM-7 was found to secrete histamine and result in the wheal and flare reactions.

4.1 β-Casomorphins and Gastrointestinal Regulation

The proteins derived from the diet are conventionally recognized to provide energy in addition to the pool of amino acids significant for growth and maintenance of numerous body activities. Additionally, these proteins provide sensory and physico-chemical attributes of protein-rich nutrients. More recently these proteins have gained more importance while knowing the role of bioactive peptides derived from these biomolecules on digestion. These bioactivities are latent in the intact proteins and are activated only after their release during the digestion. These bioactive peptides then regulate many physiologically processes in the living organisms. The primary and secondary structures of cow milk proteins are explained extensively in literature and the peptides released from these proteins are presently areas of great interest and significance worldwide. There is enough literature available that have correlated consumption of these proteins, release of bioactive peptides and regulation of numerous physiological processes in animals and humans. These properties may be related to the overall health status or decreased risk of some chronic complications.

© Springer Nature Singapore Pte Ltd. 2020
M. R. Ul Haq, *β-Casomorphins*, https://doi.org/10.1007/978-981-15-3457-7_4

The peptides from food may perform significant role in the gastrointestinal system before proteolysis to subsequent amino acids and consequent absorption. These functions may include the regulation of digestion through gastrointestinal enzymes and variation in the nutrient absorption in the gastrointestinal system.

Many of the proteins derived from diet have demonstrated regulatory activities in the metabolic pathways or cycles and digestive processes. These biomolecules may control the mucosal system and help in release of hormones related to gastrointestinal system like secretin, cholecystokinin, gastrin and gastrin inhibitory peptide (GIP) with consequent modifications on secretions of gastric and pancreatic system and on gastrointestinal motility. These activities are partially facilitated through the amino acids present in the lumen or mucosa of the intestine immediately after digestion. These effects are also possible through the food-derived small peptides that show the activity in the intestinal lumen or on the specific targets like receptors after absorption (Daniel et al. 1990). In this context, the major milk protein casein has been reported to proficiently encourage the secretion of mucous in the intestine. This demonstrates the formation of bioactive peptides from the casein in the gastrointestinal tract and regulation of secretory processes (Daniel et al. 1990; Yvon et al. 1994; Mahe et al. 1996). One of these bioactive peptides derived from κ-casein known as caseinomacropeptide has already been reported to have an active role in this module (Yvon et al. 1994; Beucher et al. 1994; Pedersen et al. 2000). An important class of bioactive peptides found in the gastrointestinal tract after proteolysis of β-casein of milk protein is β-casomorphins (BCMs). These milk-derived peptides are found to have significant role in the lumen and mucosa of intestine. These peptides have been well-established as opioids resembling morphine in structure and activities. Therefore, these physiologically active peptides have been found to bind with mu (μ)-receptors of the gastrointestinal tract and demonstrate role in gastrointestinal motility in adults and neonates (Baldi et al. 2005). For example, the peptides derived from αS2-casein of milk protein from amino acid 165–203 known as casocidin I and released through the action of chymosin has been found to inhibit the growth of bacteria like *Staphylococcus carnosus* and *Escherichia coli* (Zucht et al. 1995).

BCMs have depicted a role in the modulation of transport of amino acids in the intestine (Brandsch et al. 1994). Further these bioactive peptides were found to change the secretion of mucus in the intestine mucus secreting cells (Claustre et al. 2002; Trompette et al. 2003; Zoghbi et al. 2006). When these peptides derived from milk proteins were orally administered, these were found to affect the postprandial metabolism by activating pancreatic polypeptide, somatostatin and insulin secretion (Morley et al. 1988; Schusdziara 1983a, b; Takahashi et al. 1997; Froetschel 1996). These peptides have been found to reduce the suppression of intake of fats through the hormone enterostatin (White et al. 2000). Moreover, these peptides have depicted to increase the gastrointestinal transit time (Daniel et al. 1990; Becker et al. 1990; Defilippi et al. 1995; Mihatsch et al. 2005). Further, BCMs have modulated the absorption of electrolytes and water and also employ an anti-diarrheal activity in humans as well as in animals (Daniel et al. 1990). More recently a review composed by Liu and Udenigwe (2019) also highlighted the role of opioid peptides in gastrointestinal and central nervous systems (Liu and Udenigwe). Opioid peptides from milk like BCMs interact with the physiological system through opiate receptors located in central nervous system, gastrointestinal tract and some immune system. In this perspective, administration of these exogenous peptides may delay in gastric emptying and help in the inhibition of transit time of the gut of the digested products in some species (Shook and Burks 1986; Daniel et al. 1990). This indicates that these peptides have an ability to induce satiety. When morphiceptin and BCM-5 were administered from the mucosal side of the ileum (isolated from rabbit), these produced a reduced short-circuit and the effect reversed on administration of opioid antagonist, naloxone. This further implies that these peptides may influence the electrolyte transport in the intestine (Hautefeuille et al. 1986). The peptone from bovine casein and BCMs were found to stimulate the release of hormones (somatostatin, insulin and pancreatic polypeptide) in experimental animals (dogs) (Schusdziara 1983a, b). Further, BCM-5 and BCM-7 have been demonstrated to encourage the consumption of fatty diet, when these peptides were administered through intraperitoneal route or ICV (centrally) to rats (satiated) (Lin et al. 1998). Additionally Christy and coworkers in 2000 observed that BCM-7 obstructed the inhibitory influence of enterostatin on fatty diet consumption in rats fasted overnight (Christy et al. 2000). More recently, Chang et al. (2019) have found that BCM enhances the deposition of fat in chickens (broiler) through regulation of genes involved in lipid metabolism (Chang et al. 2019). There are many evidences available that have correlated these peptides with elevated net water and electrolyte absorption. Further, this association has been regarded as one of the determining factor for antidiarrheal actions that could be mediated through opioid receptors (sub-epithelial) and particular luminal binding sites on brush border membrane (Tome et al. 1987; Mansour et al. 1988; Mahé et al. 1989). When BCM-7 was injected into animals, the investigators found that this peptide slowed down the motility of gut and the effect was like that of morphine (Shah 2000). The results of this study were further in agreement with that of Andiran et al. (2003) who performed the experiments in humans. The investigators found that the infants fed with cow milk-derived infant formula developed constipation and further experienced anal fistulas. They argued that on consumption of cow milk, BCM-7 is released that binds the opiate receptors in gut and slowed down gastrointestinal motility. Feeding infants with the mother's milk proved out to be beneficial and preventive for anal fistulas, however, the consumption of cow milk had stimulated these effects (Andiran et al. 2003).

Daniel and coworkers in 1990 analyzed the effects of milk based casein and whey commercially synthesized BCMs on the gastrointestinal motility in rats using noninvasive methods employing ^{141}Ce (non-absorbable marker). The two protein suspensions of milk were labeled with this element and were given to rats via gastric tube. In comparison with casein protein suspension, the gastrointestinal transit time (GITT) and Gastric emptying rate (GER) of the tracer were smaller on feeding whey protein suspension. The difference in the two groups were entirely (GITT) or incompletely (GER) eliminated by pretreating the rats with the chemical naloxone (antagonist, opiate-receptor). Therefore, the authors reported that during digestion of milk casein, opioid peptides are produced that may slow down the gastrointestinal motility through opioid receptors. The hypothesis was confirmed on feeding animals with commercially synthesized BCM-4, D-Ala substituted BCM-4 (D-Ala-BCM-4) and BCM-5 (D-Ala- BCM-5). The authors reported that there were non-significant changes on GITT on feeding BCM-4. However, the substituted BCM-4 and BCM-5 depicted significant change (slowed down) the GITT as these BCMs demonstrate greater opioid activity and are more resistant to enzymatic hydrolytic digestion (Daniel et al. 1990).

In 2014, Barnett and coworkers studied the effects of consumption of A1 and A2 β-casein on the transit time of gastrointestinal system, the activity or regulation of brush border enzymes like dipeptidyl peptidase-4 (DPPIV), and inflammatory response in Wistar rats. The authors found a remarkable inhibition in the gastric transit and the up-regulation of DDPIV in the intestine. Further, results depicted an increase in the myeloperoxidase activity and the inflammatory marker (Barnett et al. 2014). More recently in 2016 a group of investigators from china compared the properties of A1 and A2 β-casein on inflammatory response, post-dairy digestive discomfort symptoms (PD3) cognitive processing in individuals reporting self-lactose intolerance. The study included 45 Han Chinese individuals in a double-blind that were randomized into 2 × 2 crossover trials. These individuals took milk with both the β-casein variants (A1/A2). The treatment continued for 14 days and included the parameters like PD3, the gastrointestinal activity (assessed though smart pill), the Subtle Cognitive Impairment Test (SCIT), the serum or fecal bio-markers and adverse events. The authors found that the milk having both variants of β-casein (A1 + A2) compared to milk with only A2 β-casein showed many symptoms. A remarkable elevated PD3 symptoms, the higher levels of inflammatory markers, increased gastrointestinal transit times, the decreased levels of short-chain fatty acids (SCFA) and enhanced response time and error rate on the SCIT. The intake of milk (A1 + A2) was related to deterioration of PD3 symptoms compared to baseline in lactose tolerant and lactose intolerant subjects. The intake of A2 milk couldn't exacerbate the PD3 symptoms compared to baseline in lactose tolerant and intolerant subjects. Therefore, the authors concluded that the intake of A1 milk was related to elevated inflammatory response of gastrointestinal system, deterioration of PD3 symptoms, the delay in gastrointestinal transit, and the reduced cognitive processing speed and accuracy. Since the removal of A1 milk reduced these activities, some symptoms of lactose intolerance can rise from inflammation it generates and may be circumvented by taking A2 milk (Jianqin et al. 2016).

4.2 Respiratory Effects

Bovine BCM-5, BCM-7 and morphiceptin were administered to new born rabbits (pre-term) intracerebroventricularly. The authors found that all the three BCMs demonstrated a depression in respiratory frequency and tidal volume. The effects were concentration dependent where the BCM-7 and morphiceptin effects were similar to as shown by morphine, however, BCM-5 was found ten times more potent compared to the morphine (Hedner and Hedner 1987). When these peptides were given directly in the bloodstream, the morphiceptin showed respiratory depression, however, BCM-5 and BCM-7 couldn't observe the same effect. The respiratory effects induced on administration of BCMs were reversed with naloxone, an opioid antagonist. This process is further associated with the sudden infant death syndrome (SIDS). The latter occurs in healthy infants who are aged below 1 year. This death rests unexplained even after the deep analysis like performance of autopsy, an inspection of the sight of death and an evaluation of the clinical history (Sun et al. 2003). Possibly, few BCMs were regarded as the culprits in the induction of SIDS by Peruzzo and co-investigators in 2000 who reported an advanced development of the gap between caudate nucleus and median eminence found in rats throughout the first postnatal weeks. They suggested that due to the immature brain development in the newborn rats, it is more likely that these peptides may cross the blood-brain-barrier (BBB) and reach the central nervous system. However, there are no concrete evidences that could relate consumption of these opioid peptides with incidence of SIDS. Nevertheless, the resemblances between the two syndromes (hyper endorphin syndrome and near-miss SIDS) have led to assumption that increased opioid activity (endogenous) may be one of the reasons of SIDS (Kuich and Zimmerman 1981). Therefore, in conclusion, the addition of this evidence may be that the enhanced exogenous opioid activity of BCMs released from cow milk β-casein could prompt or complement to the endogenous opioids in impairing breathing in the neonate.

4.3 Analgesic Effects

When BCMs were administered in systemic circulation (ability to bind mu-receptors in the central nervous system), these peptides were found to demonstrate sedative activity in addition to analgesic effects (Paroli 1988). BCM-4, BCM-5 and BCM-6 demonstrated this analgesic effect for a period of 45 min, BCM-7 established the same effect for 90 min or more (Matthies et al. 1984). When morphiceptin was administered in rat neonates, it resulted in the hyperanlgesic effect for about four and half month (for latter life) (Zadina et al. 1987). However, the confirmation and validation of these findings for the humans needs more research on cellular and molecular levels with signal cascade mechanisms as there are great variations in the developmental stages of neonates of humans (Fitzgerald 1987).

4.4 Tolerance and Dependence

As aforementioned repeatedly, BCMs are exogenous peptides derived from cow β-casein of milk and show properties, functions and structure quite similar to morphine. Therefore, these can be classified as the opioid peptides. In this perspective, these peptides when taken repeatedly may demonstrate the biological phenomenon of tolerance and physical dependence. The former was observed after repeated infusion of morphiceptin intrathecally (0.5 µL/h for a period of 3.5–4.5 days) in rats, though the latter (dependence) wasn't evaluated in this investigation. Nevertheless, the other group of investigators in 1983 developed a correlation between the opioid peptide, morphiceptin and physical dependence. The latter was assessed though infusion of this peptide at a dose of 1–23 µL/h for a period of 70–72 h. The peptide was given at the junction of the aqueduct and ventricle (fourth) in the experimental animals (rats). The removal was triggered through administration of the naloxone given at a dose of 4 mg/kg. In this perspective, it may be concluded that prolonged administration of BCMs may create physical dependence that is quite analogous to other opioids (Chang et al. 1983).

4.5 Cardiovascular Effects

The potent cardiovascular effects have been observed with the administration of BCMs when given through intracerebral and intravenous routes or directly to the regions of brain stem (Holaday 1983). When the opioid peptide morphiceptin was administered to the rats directly in the veins or to the cerebroventricles, they were found to produce hypotension and bradycardia (Widy-Tyszkiewicz et al. 1986; Wei et al. 1980). Nevertheless, in the rats with hypertension, the administration of morphiceptin in the ventricles demonstrated an increase in the heart rate and blood pressure and the effects were dose dependent. When isolated heart preparations from guinea pigs were incubated with BCM-5, this peptide produced a bipolar effect. At lower concentration it revealed positive ionotropic activity and at higher doses it established a cardiodepressive response (Liebmann et al. 1986).

4.6 Immunological Perspective of BCMs

Literature have clearly presented that opioid peptides like BCMs derived from food especially cow milk proteins to have roles in immunological processes including inflammation, allergy, mucus secretion, proliferation of lymphocyte and skin reactions. The suggested mechanism of these reactions is shown in Fig. 4.1.

4.6.1 Production of Mucins and Innate Immunity

Goblet cells are present in the epithelial covering that lines the intestinal and respiratory tracts. The principal function of these cells is to secrete the mucin proteins and the latter combines with other components like water to form mucus. The latter

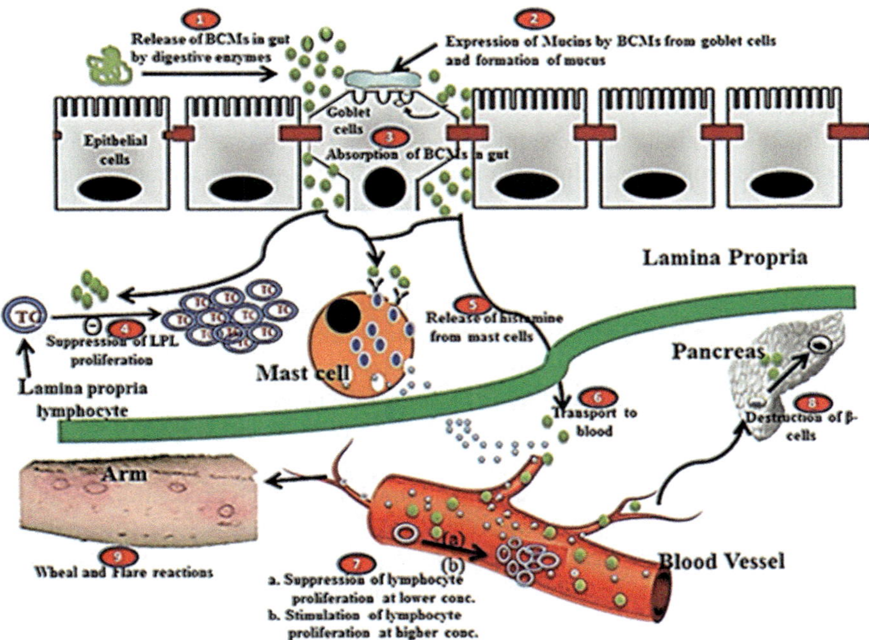

Fig. 4.1 Immunological role of BCMs on different cells

fluid serves an important barrier and hence demonstrates a significant role in the establishment of innate immunity. One of the most significant BCM i.e., BCM-7 has been observed to secrete mucin from the jejunum of rats, this effect has been reported through the activation of mu-receptor (opioid) pathway (Trompette et al. 2003). The assumption that the opioid peptides like BCM-7 bind the receptors on the mucin secreting cells (goblet cells) was analyzed in experimental animals (rat) and humans. They performed in vitro experiments on rat and human cell line in DHE and HT29-MTX respectively. In the former cell lines, the incubation with the commercially synthesized BCM-7, results depicted an increase in the expression of mucus related proteins rMuc2 and rMuc3. However, in the latter cell lines, it was found that this peptide in vitro increased the MUC5AC expression at mRNA levels and then production of mucin protein (Zoghbi et al. 2006).

From immunological perspective, the expression of mucin proteins and formation of mucus is extremely significant as it serves the innate immunity. This is more evidenced in experimental animals that the MUC2 gene that is one of the major mucin in the colon. The intestine of these animals has epithelial cells in direct contact with the bacteria. This direct exposure results in the induction of inflammation and development of cancer (Johansson et al. 2008). In this regard, it was clearly observed that commercially synthesized BCM-7 contributes considerably in the mucin secretion through the interaction of mu-opioid receptor pathway while acting on mucus producing goblet cells. Therefore, it may be assumed that the milk and other dairy products that have probability of releasing BCM-7 may have a role in the intestine immunity. This peptide may help in production and secretion of mucin and in turn formation of mucus and development of intestinal defense, maintain innate immune response and may demonstrate nutritional and health applications (Trompette et al. 2003).

This physiologically active peptide binds the mu-receptors in the epithelium and enhances the protection of intestine and hence may have associations for ameliorated intestinal strength (Zoghbi et al. 2006). The functionality of BCMs like BCM-4, BCM-6 and BCM-7 in the production of mucin was analyzed in vitro in jejunum of rats (isolated perfused). The study showed that administration of BCM-7 triggered the expression and production of mucus in the rat jejunum. This effect was assumed to occur through opiate pathway, and further, the opioid antagonist naloxone reversed the effect. This was further validated with opioid agonist (DAMGO), the administration of the latter further increased the potential release of mucus. The same effects however couldn't be observed with BCM-4 and BCM-6. Overall, the results leads to conclude that the opioid peptides derived from milk on gastrointestinal digestion may ameliorate the innate immunity through production of mucins and formation of mucus (Trompette et al. 2003).

An assumption was laid down by the researchers (Bartley and McGlashan), while considering the role of BCM-7 in the expression, production and secretion of mucins and furthers the formation of mucus and its intestinal discharge. They speculate that A1 milk that is a source of BCM-7 has the possibility to help in the production of mucus in the respiratory tract of the individuals having enhanced gut permeability (leaky gut). It is thought that BCM-7 in such individuals may bind specifically with the opiate receptors of the goblet cells (mucus producing cells) in the respiratory tract to up-regulate the expression of MUC5AC gene and in turn form more mucus and stimulation of symptoms including rhinitis and asthma (Bartley and McGlashan 2010).

There are however many considerations in this regard that include intake of A1 milk that is not favored compared to A2 milk (hypothesis), the concerns about absorption of these types of opioid peptides in the intestine, their survival in the intestine, and their interactions with other cells (immune cells) in the blood having mu-receptors. When all these concerns may be taken into account and don't interfere in this aspect, then it may be justified that why the individuals having symptoms like asthma and rhinitis may ameliorate on consumption of milk and other dairy free nutrition. The other significant parameter regarded will be the dose or concentration that will be effective in showing these symptoms.

4.6.2 Proliferation of Lymphocytes and Cellular Immunity

When an antigen in the form of protein or other foreign molecule encounters a lymphocyte in an organism, the latter start proliferating and is known as lymphocyte proliferation. The aim of the latter immunological process is to increase the number of such lymphocytes which can rapidly recognize such antigen and help in their elimination. BCMs have a significant role in cellular immunity that can modulate the mucosal immune system in humans.

This effect was analyzed in vitro by a group or researchers while incubating these opioid peptides with the lymphocytes taken from lamina propria. The authors reported suppression in such effect at lower concentration (as low as 10 mM). The activity of these peptides in showing this effect was validated by incubating these immune cells with opioid antagonist naloxone, which reversed this immunological process (Elitsur and Luk 1991). The role of these peptides in immunomodulation was also studied while incubating these peptide like BCM-7 and BCM-10 with human lymphocytes (peripheral blood), the authors found proliferation of lymphocytes at lower concentration, however, at higher doses, lymphocyte stimulation was observed (Meisel and Bockelmann 1999). Moreover, the other researchers extracted the aspirates of the intestine in human subjects, they observed the presence of BCMs in such fluid after taking cow milk (Svedberg et al. 1985).

In this perspective, Migliore-Samour and Jolles assumed that there is high specificity between these opioid peptides like BCMs and the opiate receptors present on the immune cells. These peptides exploit this binding property to either stimulate or suppress the lymphocyte proliferation. Similarly, the other investigators found that the peptides rich in proline residues and resistant to proteolytic cleavage isolated the colostrum had the bipolar functions of suppressive and stimulatory action on lymphocytes and a role in cellular immunity (Migliore-Samour and Jolles 1988). Quite evidently, the BCM peptides derived from β-casein of cow milk being rich in proline residues may also express immunomodulatory activities in cellular immunity (Zimecki et al. 1982). Moreover, more studies are required on the role of these peptides in immune cell regulation and more importantly the dosage range in which these peptides will act on such immune cells (Meisel and Bockelmann 1999).

4.6.3 Milk Allergy

Food allergy in general and milk protein allergy in particular is an immune reaction when the body's immune system erroneously recognizes this biomolecule (protein) or its derived product (peptide) as detrimental. Some of the proteins or peptides have properties like having more proline residues making them more resistant to proteolytic attack during digestion. These peptides therefore find more chance to cross the gut and more often in leaky gut conditions. These peptides like BCMs also demonstrate opioid properties and have their opiate receptors on some immune cells and therefore stimulate the B-cells the production or secretion of antibodies. The latter find more chances to bind with histamine releasing cells like mast cells or basophils. When the same antigen encounters enters, it leads to the crosslinking that in turn causes degranulation to secrete allergic compounds like histamine.

The functional significance of BCMs in the allergic reactions came into notice when it was found that morphine (agonist of BCMs) and other opioid compounds induce the secretion of histamine both in in vivo and in vitro conditions from the mast cells. These allergic reactions could be linked to the anaphylactic reactions. In this perspective, the significance of BCM-7 in showing allergic reactions was studied when this peptide was incubated with leukocytes (peripheral) from healthy human subjects. This in vitro study leads to the observation that this peptide leads to the secretion of histamine from these cells and the effect was dose dependent. The in vivo studies in humans were also performed with this peptide. This peptide was injected intradermally, and it resulted into the wheal and flare response in the skin and the allergic responses were similar to that of codeine and histamine. When the individuals were pretreated with cetirizine or terfenadine (H1 antagonist), these reactions were remarkably inhibited. The naloxone (antagonist) when injected intradermally caused an inhibition in the secretion of histamine and the skin allergic responses only about 100 times more compared to BCM-7 (Kurek et al. 1992).

Kurek and Malaczynska also investigated the role of BCM-7 in vivo in guinea pigs and in vitro in peritoneal cells derived from rats. The results showed that this peptide caused histamine release from peritoneal cells in vitro and induced wheal formation and obstruction in the bronchia in sensitized animals (guinea pigs) (Kurek and Malaczynska 1999). The opioid peptide BCM-7 was established to release the allergic molecule histamine directly in humans. This peptide was observed to induce allergic symptoms like formation of wheal and production of flare reactions. These allergic reactions resembled the same expressed by codeine and histamine and the effect was found dose dependent. Nevertheless, experimental significance of these observations warrants more studies. An association was established between the enzyme present in serum dipeptidyl peptidase IV (DPP IV) activity and the amount of BCM in the mother's milk. Two groups were formed like healthy and allergic infants. In the latter, the increased levels of BCM in the milk of mother's parallels with the lower DPP IV activity in the infant's serum. The decreased levels of BCM-7 and BCM-5 in the milk of mothers (allergic group) explains that these peptides may be transported to the blood from the intestine and demonstrate increased half-life owing to a decreased DPP IV activity (Wasilewska et al. 2011a, b). The results from our previous laboratory are also in agreement with the previous observations where a remarkable secretion of histamines was observed in addition to increased enzyme activity (tryptase specific to mast cell) on incubation of BCM-5 with cells (bone marrow derived) in vitro.

4.6.4 Milk Intolerances

Some people take A2 milk with the assumption that it becomes easy for them to digest. Nevertheless, as we know this type of milk also contains lactose in appreciable amounts. The latter is regarded as one of the most significant milk intolerance matter. The probable justification for this deceptive inconsistency may be twofold. First, the opioid peptides like BCM-7 that is produced in the gastrointestinal tract during digestion from A1β-casein of cow milk may slow down the passage of food through the intestinal tract (an attribute of other opioids also). This effect may provide enough time for the lactose to ferment. Second, some individuals may be directly intolerant to milk derived opioid peptides like BCM-7. Nevertheless the second assumption needs more research with in vitro and in vivo trials in humans with explored signal cascades.

References

Andiran F, Dayi S, Mete E (2003) Cow Milk consumption in constipation and anal fissure in infants and young children. J Paediatr Child Health 39:329–331

Baldi A, Ioannis P, Chiara P et al (2005) Biological effects of milk proteins and their peptides with emphasis on those related to the gastrointestinal ecosystem. J Dairy Res 72:66–72

Barnett MP, McNabb WC, Roy NC et al (2014) Dietary A1 β-casein affects gastrointestinal transit time, dipeptidyl peptidase-4 activity, and inflammatory status relative to A2 β-casein in Wistar rats. Int J Food Sci Nutr 65:720–727

Bartley J, McGlashan SR (2010) Does milk increase mucus production? Med Hypotheses 74:732–734

Becker A, Hempel G, Grecksch G et al (1990) Effects of beta-casomorphin derivatives on gastrointestinal transit in mice. Biomed Biochim Acta 49(11):12031–12037

Beucher S, Levenez F, Yvon M et al (1994) Effects of gastric digestive products from casein on CCK release by intestinal cells in rat. J Nutr Biochem 5(12):578–584

Brandsch M, Brust P, Neubert K et al (1994) β-Casomorphins–chemical signals of intestinal transport systems. In: Brantl V, Teschemacher H (eds) Beta-casomorphins and related peptides: recent development. VCH, Weinheim, pp 207–219

Chang KJ, Wei ET, Killian A et al (1983) Potent morphiceptin analogs: structure activity relationships and morphine like activities. J Pharmacol Exp Ther 227:403–408

Chang WH, Zheng AJ, Chen ZM et al (2019) β-Casomorphin increases fat deposition in broiler chickens by modulating expression of lipid metabolism genes. Animal 13(4):777–783. https://doi.org/10.1017/S1751731118002197

Christy L, White GA, Bray A et al (2000) Intragastric β-casomorphin-7 attenuates the suppression of fat intake by enterostatin. Peptides 21:1377–1381

Claustre J, Toumi F, Trompette A et al (2002) Effects of peptides derived from dietary proteins on mucus secretion in rat jejunum. Am J Physiol Gastrointest Liver Physiol 283(3):521–528

Daniel H, Vohwinkel M, Rehner G (1990) Effect of casein and beta-casomorphins on gastrointestinal motility in rats. J Nutr 120(2):252–257

Defilippi C, Gomez E, Charlin V et al (1995) Inhibition of small intestinal motility by casein: a role of β-casomorphins? Nutrition (Burbank, Los Angeles County, Calif) 11(6):751–754

Elitsur Y, Luk GD (1991) Beta-casomorphin (BCM) and human colonic lamina propria lymphocyte proliferation. Clin Exp Immunol 85:493–497

Fitzgerald M (1987) Pain and analgesia in neonates. Trends Neurosci 10:344–346

Froetschel MA (1996) Bioactive peptides in digesta that regulate gastrointestinal function and intake. Am Soc Anim Sci 74(10):2500–2508

Hautefeuille M, Brantl V, Dumontier AM et al (1986) In vitro effects β-casomorphins on ion transport in rabbit ileum. Am J Phys 250:G92–G97

Hedner J, Hedner T (1987) Beta-casomorphins induce apnea and irregular breathing in adult rats and newborn rabbits. Life Sci 41(20):2303–2312

Holaday JW (1983) Cardiovascular effects of endogenous opiate systems. Annu Rev Pharmacol Toxicol 23:541–549

Jianqin S, Leiming X, Lu X et al (2016) Effects of milk containing only A2 beta casein versus milk containing both A1 and A2 beta-casein proteins on gastrointestinal physiology, symptoms of discomfort, and cognitive behavior of people with self-reported intolerance to traditional cows' milk. Nutr J 15(1):45

Johansson MEV, Phillipson J, Petersson A et al (2008) The inner of the two Muc2 mucin-dependent mucus layers in colon is devoid of bacteria. Proc Acad Nat Sci 105:15064–15069

Kuich TE, Zimmerman C (1981) Could endorphins be implicated in sudden infant-death syndrome? New Eng F Med 304:973–976

Kurek M, Malaczynska T (1999) Food allergy, atopic dermatitis-nutritive casein formula elicits pseudoallergic skin reactions by prick testing. Int Arch Allergy Immunol 118(2):228–229

Kurek M, Przybilla B, Hermann K et al (1992) A naturally occurring opioid peptide from cow's milk, beta-casomorphine-7, is a direct histamine releaser in man. Int Arch Allergy Immunol 97:115–120

Liebmann C, Barth A, Neubert K (1986) Effects of β-casomorphin on 3H-ouabain binding to Guinea-pig heart membranes. Pharamzie 41:670–671

Lin L, Umahara M, York DA et al (1998) β-Casomorphins stimulate and enterostatin inhibits the intake of dietary fat in rats. Peptides 19:325–331

Liu Z, Udenigwe CC (2019) Role of food-derived opioid peptides in the central nervous and gastrointestinal systems. J Food Biochem 43(1):e12629. https://doi.org/10.1111/jfbc.12629

Mahé S, Tomé D, Dumontier AM et al (1989) Absorption of intact beta-casomorphins (beta-CM) in rabbit ileum in vitro. Reprod Nutr Dev 29(6):725–733

Mahe S, Roos N, Benamouzig R et al (1996) Gastrojejunal kinetics and the digestion of [N-15] beta-lactoglobulin and casein in humans: the influence of the nature and quantity of the protein. Am J Clin Nutr 63(4):546–552

Mansour A, Tome D, Rautureau M et al (1988) Luminal anti-secretory effects of a [beta]-casomorphin analogue on rabbit ileum treated with cholera toxin. Pediatr Res 24(6):751–755

Matthies H, Stark H, Hartrodt B (1984) Derivatives of β-casomorphins with high analgesic potency. Peptides 5:463–470

Meisel H, Bockelmann W (1999) Bioactive peptides encrypted in milk proteins: proteolytic activation and thropho-functional properties. Antonie Van Leeuwenhoek 76(1):207–215

Migliore-Samour D, Jolles P (1988) Casein, a prohormone with an immunomodulating role for the newborn? Exp Dermatol 44:188–193

Mihatsch WA, Franz AR, Kuhnt B et al (2005) Hydrolysis of casein accelerates gastrointestinal transit via reduction of opioid receptor agonists released from casein in rats. Biol Neonate 87(3):160–163

Morley JE, Levine AS, Yamada T et al (1988) Effect of exorphins on gastrointestinal function, hormonal release, and appetite. Gastroenterology 84(6):1517–1523

Paroli E (1988) Opioid peptides from food (exorphins). World Rev Nutr Diet 55:58–63

Pedersen NL, Nagain-Domaine C, Mahe S et al (2000) Caseinomacropeptide specifically stimulates exocrine pancreatic secretion in the anesthetized rat. Peptides 21(10):1527–1535

Schusdziara V (1983a) Effect of beta-casomorphins on somatostatin release in dogs. Endocrinology 112(6):1948–1951

Schusdziara V (1983b) Effect of beta-casomorphins and analogs on insulin release in dogs. Endocrinology 112(3):885–889

Shah N (2000) Effects of milk-derived bioactives: an overview. Bri J Nutr 84(1):S3–S310

Shook JE, Burks TF (1986) Peripheral mu-opioid receptors mediate small intestinal transit (sit) but not analgesia in mice. Fed Proc 45:1344–1351

Sun Z, Zhang Z, Ang X et al (2003) Relation of beta-casomorphin to apnea in sudden infant death syndrome. Peptides 24(6):937–943

Svedberg J, de Haas J, Liemenstoff G et al (1985) Demonstration of beta-casomorphin immunoreactive materials in in vitro digests of bovine milk and in small intestine contents after bovine milk ingestion in adult humans. Peptides 6:825–830

Takahashi M, Moriguchi S, Suganuma H et al (1997) Identification of casoxin C, an ileum-contracting peptide derived from bovine kappa-casein, as an agonist for C3a receptors. Peptides 18(3):329–336

Tome D, Dumontier AM, Hautefeuille M et al (1987) Opiate activity and transepithelial passage of intact beta-casomorphins in rabbit ileum. Am J Physiol Gastrointest Liver Physiol 253(6):737–744

Trompette A, Claustre J, Caillon F et al (2003) Milk bioactive peptides and β-casomorphins induce mucus release in rat jejunum. J Nutr 133(11):3499–3503

Wasilewska J, Kaczmarski M, Kostyra E et al (2011a) Cow's-milk-induced infant apnoea with increased serum content of bovine β-casomorphin-5. J Pediatr Gastroenterol Nutr 52:772–775

Wasilewska J, Sienkiewicz SE, Kuzbida E et al (2011b) The exogenous opioid peptides and DPP IV serum activity in infants with apnoea expressed as apparent life threatening events (ALTE). Neuropeptides 43(3):189–195

Wei ET, Lee A, Chang JK (1980) Cardiovascular effects of peptides related to the enkephalins and β-casomorphin. Life Sci 26:1517–1522

White CL, Bray GA, York DA (2000) Intragastric beta-casomorphin (1-7) attenuates the suppression of fat intake by enterostatin. Peptides 21(9):1377–1381

Widy-Tyszkiewicz E, Czonkowski A, Szreniawski Z (1986) Cardiovascular response to morphiceptin in spontaneously hypertensive and normotensive rats. Pol J Pharmacol Pharm 38:51–56

Yvon M, Beucher S, Guilloteau P et al (1994) Effects of caseinomacropeptide (CMP) on digestion regulation. Reprod Nutr Dev 34(6):527–537

Zadina JE, Kastin AJ, Manasco PK et al (1987) Long-term hyperanlgesia induced by neonatal β-endorphin and morphiceptin is blocked by neonatal Tyr-MIF-1. Brain Res 409:10–18

Zimecki M, Janusz M, Staroscik K et al (1982) Effect of a proline-rich polypeptide on donor cells in graft-versus-host reaction. Immunology 47:141–145

Zoghbi S, Trompette A, Claustre J et al (2006) β-Casomorphin-7 regulates the secretion and expression of gastrointestinal mucins through a μ-opioid pathway. Am J Physiol Gastrointest Liver Physiol 290(6):1105–1113

Zucht HD, Raida M, Adermann K et al (1995) Casocidin-I: a casein-αs2 derived peptide exhibits antibacterial activity. FEBS Lett 372(2–3):185–188

Chapter 5
A1 Milk and Type 1 Diabetes (T1D)

Abstract Many ecological correlations have correlated consumption of A1 milk with increased risk for incidence of T1D. These correlations are supported by human case-control studies, few in vivo and in vitro studies. It is assumed that the released BCM-7 from only A1 milk is actually the hypothetical risk element. The case- control studies have depicted an increase in the antibody profile against cow β-casein in T1D. The increased antibodies or immune reactivity against β-casein has been linked with enhanced intestinal permeability, HLA haplotypes (susceptibility-linked), A1 milk derived peptides express cross-reactive antigens to the islet cell components of the pancreas, sequence homology between cow β-casein and β-cell elements and sharing of amino acids between GLUT-2 and cow milk β-casein. However, few American Nutritionists, EFSA and Truswell refute all these correlations and mechanisms.

5.1 A1 Milk and Health Concerns

Milk is the sole nutrition for infant and a part of diet for adults also. Cow milk has served this purpose for long in this regard. It has been a source of energy in addition to the functional significance in the growth and maintenance. The protein component of this diet has been a structural part in addition to a source of the biological active peptides that have a role in metabolic and other physiological regulations. Besides these remarkable significances, cow milk intake has been a subject of strong controversy about its health concerns linking with various health complications. Primarily, the researchers assumed that the milk processing techniques like homogenization and pasteurization were associated with incidence of various human illnesses (McLachlan 2001). Consequently, milk customers or clients reported that the other components that are added to this milk like antibiotics and growth hormones are linked with the incidence of these diseases (Forman 2004).

As aforementioned in previous chapters, there are two types of milk that are regarded as A1 and A2 milk. The classification is based on the type of β-casein present in the cow milk as there are at least 12 genetic variants of this protein. These are grouped into A1 and A2 "like" β-casein. The former contains β-casein variants with histidine at position 67 of this protein while the latter contains proline at this position. The presence of histidine in A1 "like" variants allows easy hydrolysis at this position during digestion to release a seven amino acid peptide, BCM-7, which isn't the case with A2 "like" variants as proline resists such cleavage at this position.

In this perspective, during the last three decades, the scenario has changed, this A1 "like" β-casein (A1 milk) rather than whole milk has been linked with the incidence of the diseases like T1D and cardio vascular diseases CVD (Elliott et al. 1999; McLachlan 2001; Virtanen et al. 2000; Laugesen and Elliott 2003; Birgisdottir et al. 2002; Monetini et al. 2002). Moreover, neural conditions including schizophrenia, autism, and sudden infant death syndrome seemed to be aggravated by consumption of this type of milk (Sun and Cade 1999; Sun et al. 1999, 2003). It is assumed that the released peptide is actually the hypothetical factor. The ecological, case-controls, few animal and in vitro trials support this hypothesis. However, there are very few controversial reports and counter-arguments in this research area. Therefore more research with human and in vivo animal trails is needed to unravel the mechanism. These mechanistic studies must be explored at the cellular and molecular levels with the use of more sophisticated and advanced techniques. If/when this hypothesis proves correct at all the mechanistic levels, then it will be mandatory for the government to formulate guidelines to lessen this type of milk in the persons who are genetically at-risk to these illnesses.

5.2   A1 Milk and Type 1 Diabetes Mellitus (T1D)

T1D, an autoimmune disorder results whenever there is a dysfunction of β-cells of pancreas necessary for insulin secretion (the enormously important cells of pancreas). This complication typically occurs in the patients that are genetically susceptible to some external factors including environmental and dietary factors (Atkinson and Eisenbarth 2001; Bluestone et al. 2010). The environment plays a significant role in the incidence of this complication via autoimmune reactions or regulation at various check points of cell cycle (Todd 1981, 2010). For example, one of the notable agents responsible includes administration of vaccines in the childhood. In this perspective, the diet or nutrition of an individual has an important role to play in the incidence of this disease. Consequently, to the best of the literature surveyed, the milk more specially cow milk that is a part of daily diet for infants and adults has been linked the incidence of T1D in genetically at-risk individuals (Chia et al. 2017; Knip et al. 2005; Coppieters and Von Herrath 2009).

Globally, the prevalence of T1D has amplified intensely and a greatest health concern to mankind (Bruno et al. 2016). The only remedy available till date is the administration of exogenous insulin; however, the other therapeutic tools are unavailable (Raskin and Mohan 2010). The reason for this unexplored mystery seems to be the multifaceted and composite nature of this health complication. Additionally, the agents that are responsible for the incidence of this disease aren't fully listed with proper cascade mechanisms. Moreover, as aforementioned, the autoimmune nature of this disorder seems to exist from the interaction between the genes (genetically at-risk) and the environmental and other dietary agents that have enough possibility to activate the immune system. The latter finds its way for the destruction of the extremely important insulin secreting β-cells of pancreatic islets. On one hand the genes that are linked with the incidence of this disorder are widely distributed even in the subjects that don't show signs of this disease in their whole life span. On the other, complete data concerning the genetic causes with this complication is scantly. Quite evidently the impact of genes (risk) is indeterminate even in the twins (homozygous), the incidence of T1D varies in them also. This, therefore, advocates for the acclamation that environmental agents have a significant role to perform in the pathogenesis of this complication (Altobelli et al. 2016). The environmental agents that add in the pathogenesis of this disorder include certain chemical, food ingredients and viral infections (Butalia et al. 2016; Virtanen 2016). The other agents that could contribute in this perspective include the abnormality with the physiology of the individuals. These may include the abnormality in the gut mucosal system, the dysbalance in the gut microbiome, biochemical abnormalities or other physiological dysfunctions. These are together grouped as the permissive gut factors responsible for the pathogenesis of T1D. The factors responsible for the pathogenesis of this complication are depicted in Fig. 5.1.

From dietary or nutritional perspective, the cow milk as such for the adults or in the form of infant formulas for infants gets paramount importance (Jeurink et al. 2013; Wiley 2012). There are reports available that are in favour and against the hypothesis of the consumption of milk and incidence of T1D; the same will be discussed comprehensively though out this chapter. Consequently literature has demonstrated that milk proteins activate immune system (auto reactive) in genetically at-risk (susceptible) human subjects or in other in vivo animal models like NOD mice (Akerblom and Vaarala 1997; Jianqin et al. 2016). The cow milk as a natural

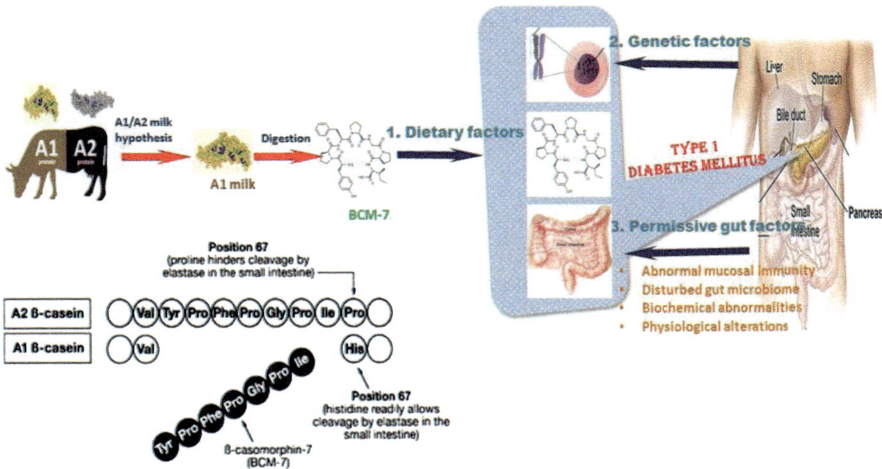

Fig. 5.1 Factors linked with pathogenesis of T1D

animal fluid is a rich source of proteins and other biomolecules in addition to defensive molecules like autoantibodies. The latter seem to mimic anti-insulin and other markers of β-cells of islets of langerhans in pancreas on consumption of milk (Harrison and Honeyman 1999). The A1 milk hypothesis regards this milk in the incidence of T1D, however, the same has been a subject of controversy by different milk companies, clients and researchers. The next discussion is on the evidences in favour and against the hypothesis.

5.3 Evidences in Favour of A1/A2 Milk Hypothesis

Through historic times, the cow milk is consumed as part of diet for adults. For infants, the cow milk may be taken as such or in the form of infant formulas. This natural animal fluid has fulfilled the purpose of nutrition in terms of energy supply to human beings. As time advanced, the composition of milk was elucidated and more sophisticated research showed other physiological activities of milk components. These bioactivities focused more on the protein fraction of milk after its digestion. The protein biomolecules after digestion yield the amino acids in the gut that are absorbed and form the amino acid pool. Some of the partially digested fragments or peptides have demonstrated their role in the biological regulation of various phenomena. However, unexpectedly this has also been a subject of controversy. The researchers initially assumed that the milk processing techniques like homogenization and pasteurization were associated with incidence of various human illnesses (McLachlan 2001). Consequently, milk customers or clients reported that the other components that are added to this milk like antibiotics and growth hormones are linked with the incidence of these diseases (Forman 2004). However, from few decades, specific milk (A1) has been regarded responsible for incidence of these health complications. The ecological, case controls, few animal and in vitro trials support this hypothesis. However, there are very few controversial reports and counter-arguments in this research area. Both the perspectives and evidences will be discussed in this chapter as follows.

5.3.1 Ecological Studies

Many ecological reports have confirmed the beneficial attributes of breastfeeding compared to consumption of cow milk in childhood (Gottlieb 2000). The effect of specified components of milk instead of whole milk gained more importance around the year 1997. These reports kept under analysis the composition of cow breeds, the polymorphism of cow milk proteins and the protein profiling in the cow breeds in different countries or regions. The A1 milk was marked as a factor responsible for pathogenesis of T1D in 1999 by Elliot and coworkers. The data was generated in children (0–14 year) from ten countries including New Zealand, USA, Australia,

Canada, Germany Denmark, Iceland, Norway, Sweden and Finland. The statistical analysis established a correlation between consumption of this type of milk and incidence of T1D. A strong association ($r = 0.726$) was observed between consumption of A1 milk and incidence of this disease. The strength of this correlation decreased on consumption of total protein ($r = 0.402$) and incidence of this disease. The association was more encouraging when A1 "like" milk (A1 and B β-casein) consumption was predominant (Elliott et al. 1999). This study was further strengthened when it was observed that in Iceland where the milk from dairy cows is predominantly A2, the incidence of T1D is also low. These ecological reports therefore laid a strong foundation in favour of this hypothesis restricting a specific type of milk for incidence of this disorder. The biochemical studies through SGID have clearly revealed release of BCM peptide from only A1 milk. This peptide has further demonstrated the opioid properties and a role in immunomodulation. Literature has clearly depicted the in vitro immunosuppressive role of this peptide on human intestinal lymphocytes (Elitsur and Luk 1991; Teschemacher 2003). Consequently, it may be assumed that the immunosuppressive nature of this peptide may create the gut immune tolerance and reduce the protection against various enteroviruses and contribute as a permissive factor for incidence of T1D (Graves et al. 1997). Thorsdottir and coworkers in 2000 performed a comprehensive ecological study in Iceland and Scandinavia while keeping accurate regards of milk protein polymorphism, status of A1/A2 milk and breed composition. The study found a low frequency of A1 milk (A1 and B) in Iceland compared to Scandinavia. Consequently the incidence of T1 in the former was also in agreement (lower) in the former countries compared to later. Therefore, the authors assumed that the lower incidence of T1D in Iceland may be justified with lower A1 allele frequency of cow milk β-casein. A more comprehensive investigation was performed by McLachlan in 2001 around sixteen countries and the data was related to children aged below 16 years. The researcher obtained the best association ($r = 0.73$) for incidence of T1D and intake of A1 milk. When A1 milk consumption was associated with B component the strength of the correlation decreased that is against Elliot et al. (1999). Nevertheless, total milk consumption correlated weakly ($r = 0.23$) with incidence of this disease (McLachlan 2001). The earlier reports were further strengthened when the data between milk protein polymorphism was correlated with the incidence of this disease. The authors found a strong correlation between the two factors in Nordic Countries and the strength of the same associationship decreased in the Iceland, where the milk is predominantly of A2 type (Birgisdottir et al. 2002). Laugesen and Elliott in 2003 observed a strong association ($r = 0.92$) with A1 milk and cream consumption and incidence of T1D. The correlation weakened on intake of A1 and B milk followed by consumption of A1, B and C milk. The assumed favored A2 milk was significantly positively associated with incidence of T1D (Laugesen and Elliott 2003). Although these studies strongly support A1 milk hypothesis, nevertheless, these are unable to justify causality. Truswell while citing the literature of Crawford et al. (2001) and Hill (2002) criticizes this hypothesis and pinpoints many limitation and lacunas. The author reports that there are many constraints in the ecological studies design and the statistical analysis. However, it must be taken to the light that the Truswell's arguments are based on the literature that are either conference papers or poster demonstrations that are weak justification or evidences compared to much comprehensive and statistically approved ecological data (Crawford et al. 2001; Hill 2002). Moreover, these researchers were linked to Fonterra that was against the hypothesis. Moreover, the data consulted by the author focused on whole dairy consumption instead of focusing on milk protein polymorphism like different variants of β-casein (A1, A2 and B). On the other hand the ecological data generated provided accurate regards of milk protein polymorphism, countries or regions and composition of breeds. Therefore, it is unjustified to provide more weightage to the studies reporting total milk intake compared to milk consumption with known polymorphisms. Besides this, it is very cumbersome to confirm the results of the data presented in posters or conferences as it lacked the information like breed composition and milk protein polymorphism. The ecological data is reliable due to the fact that there is strict assortment of regions or countries for analysis so as to decrease the confusing errors, the data were recorded carefully and finally the substantial supremacy of the statistical correlations. Overall, to the best of our comprehension, counter arguments deserve more validation, the hypothesis is therefore potentially important. As Swinburn (2004, Lay Report p 2) stated: 'it should be taken seriously''.

5.3.2 Animal Studies

Elliot and Martin in 1984 performed an animal trial in Biobreed (BB) rats in order to develop a relationship between cow milk intake and diabetes incidence (Elliot and Martin 1984). The actual risk factor for this disorder was the peptide derived from A1 milk, this peptide was named as BCM-7. Elliott and Hill regarded this peptide as hypothetical risk factor explaining the variants of β-casein as "Jeckyl and Hyde". Elliot and coworkers in 1997 studied the rates of incidence of diabetes and various diet formulations in the mice models (non-obese diabetic mice, NOD). The diet formulations were 37% chow, 0% Prosobee, 12% (Prosobee and 10% 'Samoan diet') and 36% (Prosobee and 10% 'Finnish diet'). The results indicated a considerable effect on the rate of incidence of diabetes with the diet formulation supplied with A1 β-casein (47%) compare to A2 β-casein (0%). A remarkable reduction in the incidence of diabetes was observed (28 to 2%) with casein hydrolysate. Moreover, this pathological effect was reversed on administration of BCM-7 antagonist (naloxone) in mice models (Elliot et al. 1997). Although these observations regard A1 milk as a promoter for incidence of diabetes. Nevertheless, this experimental trial had lacuna in terms of not providing the accurate diet composition. Country specific animal trials were performed in three countries with enough dairy potential including New Zealand, UK and Canada. Unfortunately, the animals used in the experiments in New Zealand got infected and hence no proper conclusions were drawn. Quite expectedly, the A1 milk was regarded more diabetogenic compared to A2 milk in the mice models of Canada. However, more unexpectedly, the

A2 milk was also found to demonstrate diabetic effects in the UK mice models. Therefore, both the variants of β-casein showed a protective role against diabetes incidence in comparison with control diets and the effect varied marginally in diabetes-inducing capacity (Beales et al. 2002). The unanticipated observations in these country-specific experiments may be due to the diet composition that was not publicized in these experiments. Moreover, the diet used in these trials included wheat that is a source of derived peptides known as gluteomorphins. The latter demonstrate opioid activity that resembles BCMs in action and could account for diabetogenic activity. In addition to this, one of the components used in these diet formulations included pregestimil that may be a source of BCMs (Cone 2002; Australian Broadcasting Corporation 2003). The nonexistence of true control may further confirm the inconsequential testing of A1 and A2 milk for their diabetogenic effect. The release of BCM-7 and other diabetogenic elements released from wheat may have definitely reproduced in these observations. While considering the limitations of these experiments, the information published in these articles may be misrepresentative until the diet composition is fully published and the experiments are designed with proper controls. In this context, the researchers or sponsors who are against the hypothesis through these trials may further anticipate that these trials were funded by Fonterra and the feed of animals was given by New Zealand Dairy Research. These two dairy companies were then intensely advocating against the hypothesis. More recently, a group of investigators from Australia analyzed the influence of A1 milk on the progression of diabetes in non-obese diabetic (NOD) mice. The observations revealed that exposure in the early life to A1 milk have diabetogenic effect that is generation dependent. Moreover, there was a significant reduction in subclinical insulitis and altered glucose handling. In addition to this a decrease in fraction of regulatory T cell subset was also seen in terms of CD4$^+$ CD25$^-$ FoxP3$^+$ cells. In conclusion, it was reported that consumption of A1 β-casein may alter glucose homeostasis and further may establish induction of T1D (Chia et al. 2018). Previously controversial observations were published by a group of researchers from china. The authors revealed a protective role of BCM-7 against oxidative stress (free radical-mediated) and hyperglycemia in rats (Yin et al. 2012). There are some limitations in these trials that used chemically synthesized peptide (BCM-7) and didn't use native A1 or A2 β-casein milk proteins. Moreover, accurate mice or rat models weren't used instead the diabetes was induced with the chemical (streptozotocin). In conclusion, these reports should be additionally reconfirmed and authenticated before making any open commendations for consumers.

5.3.3   Case Controls

The case control studies have depicted an increase in the antibody profile against cow β-casein in T1D. The effect was not related to separate A1 and A2 β-casein variants of cow milk (Cavallo et al. 1996). The immunological T cells were found to be sensitized on intake of β-casein in T1D. These cells may cross react with β-cells in the pancreas. Nevertheless, in this study the genetic variants of β-casein were not kept into consideration. Moreover, this study couldn't explain cause-effect relation of consumption of β-casein variants and incidence of T1D (Monetini et al. 2002). The subjects that were suffering from diabetes ($n = 287$), and their siblings ($n = 386$), their parents ($n = 477$) and healthy controls ($n = 107$) were assessed for antibody response against β-casein variants (A1 and A2) of cow milk. The results showed an increased antibody response against A1 β-casein in the subjects suffering from T1D (Padberg et al. 1999). This issue is very interesting and of great scope to discuss whether this increased immune response is linked with etiology of this disorder or is a manifestation of T1D. Moreover, longitudinal trials and animal or human feeding trials are urgently required to reveal the mechanism of pathogenesis of this disorder on consumption of A1 and A2 cow milk. These investigations may be designed with pronounced probability (genetically at-risk infants) and could contribute significantly to this hypothesis.

5.3.4   Sibling Studies

A Finnish group of researchers assessed the established the incidence of T1D in the individuals who developed this disorder (Virtanen et al. 2000). The follow-up time calculated as median was about 9.7 years between incidence of this disorder and infants exposure to cow milk in infancy. The influence of exposure of cow milk in childhood and infancy was linked with the enhanced genetic susceptibility expressed in terms of HLA-DQB1 markers of T1D risk. The maximum intake of cow milk may be diabetogenic in siblings with this disorder (Virtanen et al. 2000). The results indicated the relative risk as 5.37 when the cow milk was adjusted in the diet. Moreover, very lesser cases of this disorder were observed when the children took less than three (3) glasses of milk per day. They also didn't take cow milk before the age of 2 months. However, deeper research is needed to validate the events occurring between genetic susceptibility and cow milk consumption.

5.4   Critique of the A1 Milk Hypothesis and T1D

The correlation established between intake of A1 milk and prevalence of T1D based on the ecological between-country data had been criticized on the following reasons:

- It is not fully clear that the persons who developed diabetes were the ones who actually consume A1 β-casein.
- The data generated in human till date regarding A1 milk intake and development of diabetes is mostly related to human infants (Borch-Johnsen et al. 1984; Mayer et al. 1988). These infants take cow milk in the form of infant formulas. The total milk intake in the form of infant formulas in any country is very less. Therefore

the consumption of milk at the national level may not serve as measureable valuation of the A1 β-casein intake from these formulas. The latter are composed with higher ratio of whey compared to casein. Moreover, it is not necessary that the protein fraction used for preparation of infant formulas will always belong from the country where it is manufactured.

- The exclusion of confounding is also not possible. For example, people residing in Finland possess higher frequency of HLA haplotypes that show a diabetic susceptibility (Reijonen et al. 1991). Moreover, reports have also demonstrated that Sardinia Island depict the highest incidence of T1D in the Mediterranean areas. The emigration data and HLA types demonstrate that it is a genetically determined disorder (Muntoni and Muntoni 1999).
- Many case-control reports have correlated breast feeding negatively with T1D (Jones et al. 1998).
- There exists the environmental and socioeconomic variations between infants taking mother's milk and infant formulas, the former taking milk doesn't contain A1 milk.
- The Nutritionists have claimed the unreliability of associations between the consumption of food and development of chronic disease (FAO/WHO Expert Consultation 1998; Truswell 2002).
- The data showing the total A1 β-casein in the total casein at the national and international is unknown. Moreover, the average intake of this type of milk has never been published.
- In these studies some of the leading dairy corporations, companies and boards weren't taken into consideration. For instance, Ireland and Netherlands, or developing/emerging countries.

An editorial presented by Beaglehole and Jackson for the article published by Laugesen and Elliott in 2003 on ecological correlations point out that these type of associations are their best for establishment of the hypothesis. However, these studies should be followed by observational studies and more complex, difficult and costly clinical trials. All these studies will be the right option to make public or consumer recommendations (Beaglehole and Jackson 2003). Similarly these correlations are very difficult to interpret because of country differences. Therefore, these associations should be followed by collateral evidence (Altman 1991). Moreover, other more thoughtful constraint of these ecological studies is the process of reproducibility that is an essential component of scientific phenomenon. Furthermore, these correlations are undoubtedly very valuable, nevertheless not too adequate to draw conclusions concerning the human diet and nutrition with the human illnesses and at times may be misleading (Willett 1990). For instance, the lack of information regarding the consumption of tobacco and mortality rates due to ischaemic heart disease in Laugesen and Elliott's studies (Laugesen and Elliott 2003). Smoking is regarded as a recognized risk agent within the countries however; the changeability of other risk agents for this multifactorial disorder between countries complicates its consequences when nationwide averages are associated. The animal models that have been used for the diabetes studies in vivo are NOD mice and BioBreeding (BB) rats, the genetic elements of both the strains as strong as humans (Powers 2001).

The number of animals in these experimental groups that develop diabetes changes with the diet (weaning), however, in some cases wheat and soy in the diet was found to be diabetogenic (Scott 1996). On the other hand, bovine milk proteins, skim milk or casein proteins in the diet were found to induce diabetes in these model animals (Paxson et al. 1997). Nevertheless, two bovine proteins i.e., immunoglobins G and serum albumin were depicted to encourage the damage to islet cells but couldn't induce diabetes in the experimental animals (Paxson et al. 1997). In this perspective it was quite essential to validate the animal experiments performed by Elliot and coworkers in 1997. An animal experiment separately at three places (Ottawa in London, UK and University of Auckland) was carried out to confirm the same investigation. Animal experiments carried out in London showed the highest incidence of diabetes in the control group animals that were fed with wheat, soy, corn, oats, fish meal, alfalfa, and cellulose but not milk. Similarly in Ottawa the mixed cereal-based control diet showed remarkably highest incidence of diabetes. The difference in the incidence of diabetes on feeding A1 and A2 β-caseins was little and non-significant. The animal trial performed in Auckland was stopped due to the infection outbreak in animals. However, till the time of infection outbreak the results were in agreement with the two other places. There was no statistical difference in the incidence of diabetes in the animals with A1 and A2 β-casein. These investigators of the study therefore concluded that the hypothesis of A1 β-casein to be more diabetogenic was not confirmed. Additionally, the authors concluded that diabetes may be prevented by either changing or removing the milk element in the diet and they even focused more on the wheat component (Beales et al. 2002). The addition of complete milk protein or serum albumin couldn't influence the incidence of diabetes in experimental animals (BB rats) (Virtanen et al. 1991; Malkani et al. 1997). Moreover, adding the skim milk to the formulated diet also couldn't induce the diabetes in experimental animals (NOD mice) (Coleman et al. 1990; Virtanen et al. 1991).

The case-control studies in humans have testified the histories of feeding in the childhood or infancy in case of diabetes and related with control human beings corresponded with sex, age and social status. Since the T1D initiates in the childhood, the details regarding the feeding in infancy has to count on the memory of the mothers that too between 10–15 years. Thirteen case studies were reviewed and then meta-analysis was performed on the studies with the best proposals. It was concluded that the analysis with a minimum of the possible biasness, the risk for the incidence of T1D was approximately one and a half times enhances in the infants that had a history for the allergy with cow milk or fewer than 90 days breast feeding (Gerstein 1994). In a similar attempt, Norris and Scott in 1996 performed the same type of meta-analysis on seven studies. Moreover, they stressed on the possible biasness from various response rates of the control and case groups. They found a ratio (odd) for T1D and were only about 1.13 for those that were not breast-fed. They finally concluded that the enhanced risk of T1D linked with any of the infant diet exposures is minimal (Norris and Scott 1996). Similarly 253 children were analyzed from the families having T1D for three antibodies in their serum against β-cells of islets of langerhans i.e., the possible period ahead of clinical diabetes. In about all the cases (18) cases, infant feeding was statistically non-significant from the controls (Norris et al. 1996). Virtanen and coworkers in 2000 collected about 36 children, those who developed diabetes and compared with controls, all corresponded with HLA typing for inherited vulnerability to T1D. The authors found the results in agreement with the aforementioned studies whereby the children who were breast-fed for 2 months didn't differ statistically (Virtanen et al. 2000). The same authors earlier found the breast feeding protective against the T1D (Virtanen et al. 1991). The countries where these types of studies (case-control) have been performed demonstrate high A1 β-casein intake. In genetically susceptible infant for T1D, the associated risk may be due the reduction in the immunity (lack of breast feeding) instead of the consumption of milk. Goldberg and coworkers in 2002 concluded that the incidence of T1D with the consumption of A1 milk is unlikely and misleading. The soy and wheat proteins are even more diabetogenic compared to milk (Goldberg et al. 2002; Scott 1995). Researcher also reported that there are not enough studies available that clearly depict the release of BCM-7 form A1 "like" β-casein variants in humans (Hill 2002; Teschemacher et al. 1986). It was suggested by Elliott and coworkers in 1997 that this peptide (BCM-7) acts on the immune cells (lymphocytes) present in the intestinal wall and through an unknown mechanism stimulates an auto-immune response to β-cells (insulin-producing) that results in their destruction. This leads to a decrease in the secretion of insulin from these cells. Nevertheless, other researchers couldn't find that this peptide is produced in the intestines in animal trials (lambs) (Hartwig et al. 1997).

5.5   European Food Safety Authority (EFSA) Report

The effect of intake of A1 milk and its associated links with the incidence of T1D in genetically at-risk individuals is a matter of controversy. As aforementioned, the ecological, case-controls, in vivo and in vitro studies have supported the A1 milk hypothesis. On the other hand, European Food Safety Authority (EFSA) published a report in 2009 and reported that there isn't adequate data to link intake of A1 milk with the incidence of T1D (EFSA 2009). They argued that the data generated through animal trials provide contradictory observations. The ecological data is unable to develop a cause-effect associationship satisfactorily. It is also stated that the relationship between intake of A1 milk with T1D is merely a suggestive substantiation. They reported that the intake of A1 milk at national level shouldn't be the same as infant formulas. In few regions like Switzerland, the incidence of T1D has enhanced three times since 1990 without an alteration in the consumption of milk protein (Crawford et al. 2001). Nevertheless, the report highlights that these observations are adequate to formulate a concrete hypothesis.

5.6   Possible Mechanism

Many ecological and case-control studies have established an association between A1 cow milk intake and incidence of T1D. Animal trials performed in mice and rats have further supported this correlation. In vitro studies on cell lines with commercially synthesized peptide (BCM-7) have added to this hypothesis. The biochemical studies elucidating the release of BCM-7 from only A1 milk and the reports linking the incidence of T1D with this peptide in genetically at-risk subjects further back this associationship. The opioid and binding properties of this peptide resembling that of morphine revealed through pharmacological investigation support this hypothesis. The immunological findings observed through enhanced antibody profile in NOD mice and BB rats against A1 milk further establishes a role of this type of protein in the incidence of T1D. BCM-7 with opioid properties released from A1 milk decreases the lymphocyte proliferation derived from human intestine in vitro. Therefore there is enough probability that such an immunosuppression may affect the progression of gut-linked immune tolerance and in turn subdues the defense against the enteroviruses (Laugesen and Elliott 2003; Elliott et al. 1997). The possible mechanism is depicted in Table 5.1. In our laboratory, we found an inflammatory response of A1 β-casein and commercially synthesized BCM-7 and BCM-5 in mice models. We observed an increase expression of inflammatory markers like myeloperoxidase, monocyte chemotactic protein, interleukin-4 and production of histamine. Moreover an enhanced antibody profile was observed through production of IgE, IgG, IgG1/IgG2a and associated expression of TLR-2 and TLR-4 and leucocyte infiltration in mice intestine. However, there was no change in the sIgA, IgA$^+$ and goblet cell numbers. These observations demonstrate that intake of A1 β-casein and BCM-7 and BCM-5 induce inflammatory immune response in mice intestine most likely through Th2 pathway (Raies et al. 2014a, b). These immunological activities and pharmacological properties of BCM-7 may considerably be

Table 5.1 Proposed mechanisms for incidence of T1D and cow β-casein consumption

Mechanism	Explanations
Increased antibodies against β-casein or enhanced immune reactivity towards β-casein (Laugesen and Elliott, 2003)	• Enhanced intestinal permeability
	• HLA haplotypes (susceptibility-linked)
	• The A1 milk derived peptides express cross-reactive antigens to the islet cell components of the pancreas, thus activating an autoimmune reaction by the lymphocytes
	• Sequence homology between cow β-casein and β-cell elements including p 69 carboxypeptidase and GLUT-2 that serve as auto antigens in T1D
	• GLUT-2 specific to β-cell shares five successive amino acids with cow milk β-casein
A1 but not A2 variant of cow milk β-casein produces BCMs (Elliott et al. 1999)	• BCMs in general and BCM-7 in particular may suppress lymphocyte proliferation derived from human intestine and in turn suppress the progression of intestine-linked immune tolerance, or resistance against enteroviruses.

concerned in the etiology of T1D. Systematically, the perception of "cross-reactivity" between the cow milk protein derived peptides and glucose transporter (GLUT2) is appealing (Harrison and Honeyman 1999). The concept of the cow milk proteins and derived peptides in the aetiology of the incidence of T1D in genetically at-risk patients is interesting and intriguing. The exact mechanism through which A1 β-casein influences immune system is not clearly explored. The increased antibody profiles against A1 β-casein are cross-sectional studies and deserve more investigations. The additional human findings are case-control reports of immune reactions and these may be fairly in favour of A1 milk hypothesis. The instantaneous and inevitable consequence is that more investigations are required to explore the A1 milk hypothesis. Animal and human feeding trials with the exploration of immunological, biological and biochemical parameters are needed to explore the mechanism. The more advanced techniques like mass spectroscopy, ELISA, western blots, flow cytometry, histology should be exploited in this research area. A cause-effect relationship should be established from the data generated from in vivo and in vitro experiments. If/when the role of A1 β-casein is developed with the incidence of T1D, then the governments have to make final recommendations for the consumers.

5.7 Present Status of A1 Milk Hypothesis

- Supported by various ecological studies, case-control reports, animal and few human feeding trials regarding intake of A1 milk and incidence of T1D.
- Supported by in vitro trials involving incubation of commercially synthesized BCM-7 and expression of various immunologically active molecules in cell lines.
- On the other hand, EFSA report (2009) couldn't find adequate data to link intake of A1 milk with the incidence of T1D.
- Some commentators argued against the A1 cow milk hypothesis. The authors couldn't substantial or credible evidence between consumption of A1 cow milk and adverse consequences in humans.
- The matter has to be resolved through the deeper research at all the cellular, molecular, biochemical, physiological and immunological levels through exploration of the signal transduction cascade pathways.

5.8 Conclusions

The correlation between intake of A1 milk and incidence of T1D in genetically at-risk subjects is fascinating for consumer health. This hypothesis is backed by ecological and case-control studies. The animal feeding experiments performed in mice and rats are further in agreement with the hypothesis. These trials have been performed with either purified or extracted proteins or chemically synthesized peptides. The opioid nature of these peptides and their pharmacological properties further add to this hypothesis. The biochemical mechanism of BCM-7 release from only one variant further support this hypothesis. The immunological activities depicted by these peptides in cell lines are supportive of this hypothesis. On one hand the evidence in favour of the hypothesis are very strong but on the other hand the controversial arguments and cross-sectional data deserve more research. In conclusion, the current status of the hypothesis states that more research is needed in cell line to unravel the cascade mechanism. Till the hypothesis is proven correct at all the molecular, cellular and immunological levels, it the individual choice of the subjects at risk (genetically) to T1D either to eliminate this β-casein variant from the diet or to enhance the favorable variant in their diet.

5.9   Future Perspectives of A1 Milk Hypothesis

To the best of our knowledge, literature surveyed and comprehension, the present status of the A1 milk hypothesis lays onward the resulting essential actions:

- The governments of all the countries should provide more funding for the research on human and animals feeding trials to establish a cause-effect correlation between consumption of A1 cow milk and incidence of T1D.
- More and more in vitro trials should be carried out on cell lines to elucidate the mechanism of effect A1 milk and related peptides at the cellular, biochemical and molecular levels.
- The physiological and immunological perspectives should be explored between consumption of A1 cow milk and incidence of T1D.
- Through various scientific awareness programmes like seminars, workshops and conferences, the milk clients, consumers and other stakeholders should be aware off the current status of A1 cow milk hypothesis.

"If/when the A1 milk hypothesis proves correct, and then Governments will have to come up with the policies, recommendations and regulations regarding A1 cow milk hypothesis and human health. Till then, it is the individual choice of the subjects (genetically at-risk) to T1D to remove or lessen this variant of β-casein in their diet as a precautionary measure. Nevertheless, they should adopt this strategy knowing that there is considerable improbability about the benefits of assuming such an approach."

References

Akerblom HK, Vaarala O (1997) Cow Milk proteins, autoimmunity and type 1 diabetes. Exp Clin Endocrinol Diabetes 105(2):83–85
Altman DG (1991) Practical statistics for medical research. Chapman and Hall, London, p 298

Altobelli E, Petrocelli R, Verrotti A et al (2016) Genetic and environmental factors affect the onset of type 1 diabetes mellitus. Pediatr Diabetes 17(8):559–566

Atkinson MA, Eisenbarth GS (2001) Type 1 diabetes: new perspectives on disease pathogenesis and treatment. Lancet 358:221–229

Australian Broadcasting Corporation (2003) White mischief transcript. www.abc.net.au/4corners/content/2003/transcripts/s820943.htm

Beaglehole R, Jackson R (2003) Balancing research for new risk factors and action for the prevention of chronic diseases. N Z Med J 116:291–292

Beales PE, Elliott RB, Flohé S (2002) A multicenter, blinded international trial of the effect of A1 and A2 beta casein variants on diabetes incidence in two rodent models of type 1 diabetes. Diabetologia 45(9):1240–1246

Birgisdottir BE, Hill JP, Harris DP et al (2002) Variation in consumption of cow milk proteins and lower incidence of type 1 diabetes in Iceland vs. the other 4 Nordic countries. Diabetes Nutr Metab 15:240–245

Bluestone JA, Herold K, Eisenbarth G (2010) Genetics, pathogenesis and clinical interventions in type 1 diabetes. Nature 464:1293–1300

Borch-Johnsen K, Joner G, Mandrup-Poulsen T, Christy M, Zachau-Christiansen B, Kastrup K, Nerup J (1984) Relation between breast-feeding and incidence rates of insulin dependent diabetes mellitus. A hypothesis. Lancet 2:1083–1086

Bruno G, Pagano E, Rossi E et al (2016) Incidence, prevalence, costs and quality of care of type 1 diabetes in Italy, age 0-29 years: the population-based CINECA-SID ARNO observatory, 2002-2012. Nutr Metab Cardiovasc Dis 26(12):1104–1111

Butalia S, Kaplan GG, Khokhar B et al (2016) Environmental risk factors and type 1 diabetes: past, present, and future. Can J Diabetes 40(6):586–593

Cavallo MG, Fava D, Monetini L et al (1996) Cell-mediated immune response to beta casein in recent-onset insulin-dependent diabetes: implications for disease pathogenesis. Lancet 348(9032):926–928

Chia JSJ, McRae JL, Kukuljan S et al (2017) A1 beta-casein milk protein and other environmental pre-disposing factors for type 1 diabetes. Nutr Diabetes 7(5):e274. https://doi.org/10.1038/nutd.2017.16

Chia JSJ, McRae JL, Enjapoori AK et al (2018) Dietary cows' milk protein a1 beta-casein increases the incidence of T1D in NOD mice. Nutrients 10(9):E1291. https://doi.org/10.3390/nu10091291

Coleman DL, Kazuva JE, Leiter EH (1990) Effect of diet on incidence of diabetes in non-obese diabetic mice. Diabetes 39:432–436

Cone DH (2002) Secret memo reveals Fonterra alarm. Nat Bus Rev p 14

Coppieters KT, Von Herrath MG (2009) Histopathology of type 1 diabetes: old paradigms and new insights. Rev Diabet Stud 6(2):85–96

Crawford RA, Boland MJ, Norris CS et al (2001) Milk containing beta-casein with proline at position 67 does not aggravate neurological disorders. Application PCT/NZ2001/000186 events

Elitsur Y, Luk GD (1991) Beta-casomorphin (BCM) and human colonic lamina propria lymphocyte proliferation. Clin Exp Immunol 85:493–497

Elliot RT, Martin JM (1984) Dietary protein: a trigger of insulin dependent diabetes in BB rat. Diabetologia 26(4):297–299

Elliott RB, Wasmuth HE, Bibby NJ et al (1997) The role of β-casein in the induction of insulin-dependent diabetes in the non-obese diabetic mouse and humans. Int Dairy Bull 9702(SI):445–453

Elliott RB, Harris DP, Hill JP et al (1999) Type 1 (insulin-dependent) diabetes mellitus and cow milk: casein variant consumption. Diabetologia 42:292–296

European Food Safety Authority (EFSA) (2009) Review of the potential health impact of β-casomorphins and related peptides. Sci Rep 231:1–107

FAO/WHO Expert Consultation (1998) Carbohydrates in human nutrition. Report of a Joint FAO/WHO Expert Consultation, FAO Food and Nutrition Paper 66. Food and Agriculture Organization of the UN, Rome

Forman JQ (2004) Should I be worried about bovine growth hormones in milk? Globe, Boston

Gerstein HC (1994) Cow's milk exposure and type 1 diabetes mellitus. A critical review of the clinical literature. Diabetes Care 17:13–19

Goldberg JP, Folta C, Must A (2002) Milk: can a 'good' food be so bad? Pediatrics 110:826–832

Gottlieb S (2000) Early exposure to cows' milk raises risk of diabetes in high risk children. Bri Med J 321(7268):1040D

Graves PM, Pallansch MA, Noris JM et al (1997) The role of enter viral infections in the development of IDDM: limitations of current approaches. Diabetes 46:161–168

Harrison LC, Honeyman MC (1999) Cow's milk and type 1 diabetes: the real debate is about mucosal immune function. Diabetes 48(8):1501–1507

Hartwig A, Teschemacher H, Lehmann W, Gauly M, Erhardt G (1997) Influence of genetic polymorphisms in bovine milk on the occurrence of bioactive peptides. In: Milk protein polymorphism. International Dairy Federation, Brussels, pp 459–460

Hill J (2002) National Business Review, NZ-With reference to: 2001 Internal Dairy Board memorandum

Jeurink PV, Van Bergenhenegouwen J, Jiménez E et al (2013) Human milk: a source of more life than we imagine. Benefic Microbes 4(1):17–30

Jianqin S, Leiming X, Lu X et al (2016) Effects of milk containing only a2 beta-casein versus milk containing both a1 and a2 beta-casein proteins on gastrointestinal physiology, symptoms of discomfort, and cognitive behavior of people with self-reported intolerance to traditional cows' milk. Nutr J 15(1):45

Jones M, Swerdlow AJ, Gill LE, Goldacre MJ (1998) Pre-natal and early life risk factors for childhood onset diabetes mellitus: a record linkage study. J Epidemiol 27:444–449

Knip M, Veijola R, Virtanen SM et al (2005) Environmental triggers and determinants of type 1 diabetes. Diabetes 54:S125–S136

Laugesen M, Elliott R (2003) Ischaemic heart disease, type 2 diabetes, and cow milk a1 β-casein. N Z Med J 116:1–19

Malkani S, Nomplegi D, Hanse JW, Greiner DL, Mordes JP, Rossini AA (1997) Dietary cow's milk protein does not alter the frequency of diabetes in the BB rat. Diabetes 46:1133–1140

Mayer EJ, Hamman RF, Gay EC, Lezotte DC, Savitz DA, Kingensmith GJ (1988) Reduced risk of IDDM among breast-fed children. Diabetes 37:1625–1632

McLachlan CNS (2001) β-Casein A1, ischaemic heart disease mortality, and other illnesses. Med Hypothesis 56:262–272

Monetini L, Cavallo MG, Manfrini S et al (2002) Antibodies to bovine beta-casein in diabetes and other autoimmune diseases. Horm Metab Res 34:455–459

Muntoni S, Muntoni S (1999) New insights into the epidemiology of type 1 diabetes in Mediterranean countries. Diabetes Metab Res Rev 15:133–140

Norris JM, Scott FW (1996) A meta-analysis of infant diet and insulin-dependent diabetes mellitus: do biases play a role? Epidemiology 7:87–92

Norris JM, Klingensmith G, Yu L, Hoffman M, Chase HP, Erlich HA, Hamman RF, Eisenbarth GS, Rewers M (1996) Lack of association between early exposure to cow's milk protein and b cell autoimmunity. Diabetes autoimmunity study in the young (DAISY). J Am Med Assoc 276:609–614

Padberg S, Schumm-Draeger PM, Petzoldt R et al (1999) The significance of a1 and a2 antibodies against beta-casein in type-1 diabetes mellitus. Deut Med Wochenschr 124(50):1518–1521

Paxson JA, Weber JG, Kulczycki A (1997) Cow's milk-free diet does not prevent diabetes in NOD mice. Diabetes 46:1711–1717

Powers AS (2001) Diabetes mellitus. In: Braunwald E, Fauci AS, Kasper DL, Hauser SL, Longo DL, Jameson JL (eds) Harrison's principles of internal medicine, 15th edn. McGraw Hill, New York, pp 2112–2114

Raies MH, Kapila R, Sharma R et al (2014a) Comparative evaluation of cow β-casein variants (a1/a2) consumption on th2-mediated inflammatory response in mouse gut. Eur J Nutr 53(4):1039–1049

Raies MH, Kapila R, Saliganti V (2014b) Consumption of β-casomorphins-7/5 induce inflammatory immune response in mice gut through th2 pathway. J Funct Foods 8:150–160

Raskin P, Mohan A (2010) Emerging treatments for the prevention of type 1 diabetes. Expert Opin Emerg Drugs 15(2):225–236

Reijonen H, Ilonen J, Knip M, Kerblom HK (1991) HLA-DQB1 alleles and absence of Asp 57 as susceptibility factors of IDDM in Finland. Diabetes 40:1640–1644

Scott F (1995) AAP recommendations on cow milk, soy and early infant feeding. Pediatrics 96:515–517

Scott FW (1996) Food-induced type1 diabetes in the BB rat. Diabetes Metab Rev 12:341–359

Sun Z, Cade JR (1999) A peptide found in schizophrenia and autism causes behavior changes in rats. Autism 3:85–95

Sun Z, Cade JR, Fregly M (1999) β-Casomorphin induces Fos-like immunoreactivity in discrete brain regions relevant to schizophrenia and autism. Autism 3:67–83

Sun Z, Zhang Z, Wang X et al (2003) Relation of β-casomorphin to apnea in sudden infant death syndrome. Peptides 24:937–943

Swinburn B (2004) Beta-casein A1 and A2 in milk and human health. Report to New Zealand Food Safety Authority. Prepared for New Zealand Food Safety Authority

Teschemacher H (2003) Opioid receptor ligands derived from food proteins. Curr Pharm Design 9(16):1331–1344

Teschemacher H, Umbach M, Hamel U, Praetorius K, Ahert-Hilder G, Brantl V, Lottspeich F, Henschen A (1986) No evidence for the presence of b-casomorphins in human plasma after ingestion of cow's milk or milk products. J Dairy Res 53:135–138

Thorsdottir I, Birgisdottir BE, Johannsdottir IM, Harris DP, Hill J, Steingrimsdottir L, Thorsson AV (2000) Different β-casein fractions in Icelandic versus Scandinavian cow's milk may influence diabetogenicity of cow's milk in infancy and explain low incidence of insulin-dependent diabetes mellitus in Iceland. Pediatrics 106(4):719–724

Truswell AS (2002) Meat consumption and cancer of the large bowel. Eur J Clin Nutr 56(1):S19–S24

Todd JA (1981) A protective role of the environment in the development of type 1 diabetes? Diabet Med 8:906–910

Todd JA (2010) Etiology of type 1 diabetes. Immunity 32:457–467

Virtanen SM (2016) Dietary factors in the development of type 1 diabetes. Pediatr Diabetes 17(22):49–55

Virtanen SM, Laara E, Hypponen E et al (2000) Cow's milk consumption and the childhood diabetes in Finland study group, HLADQB1 genotype, and type 1 diabetes. Diabetes 49:912–917

Virtanen SM, Rasanen L, Aro A, Lindstrom J, Sipppola H, Lounama R, Toivanen L, Tuomilehto J, kerblom HK (1991) Infant feeding in children of 7 years of age with newly diagnosed IDDM. Diabetes Care 14:415–417

Wiley AS (2012) Cow milk consumption, insulin-like growth factor-i, and human biology: a life history approach. Am J Hum Biol 24(2):130–138

Willett W (1990) Nutritional epidemiology. Oxford University Press, New York, pp 8–9

Yin H, Miao J, Ma C et al (2012) β-Casomorphin-7 cause decreasing in oxidative stress and inhibiting NF-κB-iNOS-NO signal pathway in pancreas of diabetes rats. J Food Sci 77(2):C278–C282

Chapter 6
A1 Milk and Heart Diseases

Abstract Many ecological correlations have correlated consumption of A1 milk with increased risk for incidence of CHD. These correlations are supported by human case-control studies, few in vivo and in vitro studies. It is assumed that the released BCM-7 from only A1 milk is actually the hypothetical risk element. BCMs released from A1 β-casein is proposed to play role in the progression of atherogenesis. The latter is assumed to occur through the stimulation of the oxidation of LDL by BCM-7. One of the most likely justification that cow milk β-casein could be atherogenic are the in vitro findings that demonstrated the oxidation of LDL-cholesterol to oxidized LDL-cholesterol. This biochemical finding was later validated through immunological perspective, where antibodies were assessed in infants against the oxidized LDL-cholesterol. However, few American Nutritionists, EFSA and Truswell refute all these correlations and mechanisms.

6.1 Cow Milk Peptides and Cardiovascular Health

During the last few decades the role of proteins and peptides in the healthiness of cardiovascular system has gained much significance. The peptides have been studied in particular in this research area that demonstrates the antihypertensive attributes or properties of these elements. These peptides establish even more importance when these are derived from the diet and help in regulating the cardiovascular system. Moreover, these peptides have also been observed to release from the milk that is used as a source of food for infants and adults. There have been many animal studies and human reports that have established the role of these peptides in reducing the blood pressure. In this regard, a tripeptide (lactotripeptide) released from milk with an amino acid sequence (Ile-Pro-Pro) has already been found to specifically escapes the intestinal complete digestion and finds a chance to reach the blood (Foltz et al. 2007). There have been many reports available in the literature that have demonstrated the attributes and clinical significance of these antihypertensive cow milk derived peptides (FitzGerald and Meisel 2000; Hartmann and Meisel 2007; Korhonen and Pihlanto 2006; Saito 2008; Hong et al. 2008; Moller et al. 2008). The most significant mechanism elucidated through in vitro studies is the inhibition of

© Springer Nature Singapore Pte Ltd. 2020 83
M. R. Ul Haq, *β-Casomorphins*, https://doi.org/10.1007/978-981-15-3457-7_6

ACE. Nevertheless, there is enough discussion going on regarding its significance and long-term influence on the clinical studies. In this regard the regulation of blood pressure by the peptides derived from food in general and cow milk in particular is a very significant research field in nutritional biochemistry.

6.2   A1 Milk and Heart Diseases

On one hand, milk and its derived fragments have been found to have beneficial role in the regulation of blood pressure. But on the other hand, during the last few decades there has been an establishment of a correlation between consumption of a specific type of milk i.e. A1 milk and its adverse role in ischemic heart disease (IHD) (McLachlan 2001; Laugesen and Elliott 2003a, b; Tailford et al. 2003). Nevertheless, it has also been found that some of the populations like East African Masai and Northern Keyan Samburu have very less or even no progression of heart disease after consuming milk rich diets. Scientist found that the milk in these regions is A2 type collected from the Zebu cattle (McLachlan 2001) that is the favorable milk variant. Similarly, the Western countries consume the diet rich in cow milk primarily from Holstein breed and therefore demonstrate greater incidence of CVD. These cows have been observed to show greater A1 allele frequency of β-casein protein. Many epidemiological studies have already analyzed the A1 and A2 allele frequencies, the consumption of A1 milk and incidence of heart diseases (McLachlan 2001; Laugesen and Elliott 2003a, b).

6.3   LDL Oxidation

BCMs released from A1 β-casein of cow milk is proposed to have a role in the development of atherogenesis. In this perspective, one of the important and among potent BCMs, BCM-7 is suggested to play a role in the proatherogenesis (Allison and Clarke 2006). The latter is assumed to occur through the stimulation of the oxidation of one of the most significant lipoprotein that is known as low-density lipoproteins (LDL). The latter lipoprotein is one of the most important transporters of cholesterol in blood and is composed of phospholipids (PLs), cholesterol (free and ester form), triacylglycerol and the protein fraction as apolipoprotein B100 (ApoB100) (Stocker and Keaney Jr 2004). Very low density lipoprotein (VLDL) is the main source from which LDL is derived in blood. This lipoprotein is responsible for transport of cholesterol to peripheral tissues. The apoprotein ApoB100 in LDL is a single polypeptide with a molecular weight of 500 kDa and its synthesis occurs in the liver. This lipoprotein is found to be very vulnerable to oxidative damage (Steinberg 1997, 2002; Stocker and Keaney Jr 2004; Beck et al. 2008; Matsuura et al. 2008). This susceptibility of LDL to undergo oxidation is one of the key elements that are responsible for the establishment of atherosclerosis and in turn development of cardiovascular complications. In sever atherosclerotic conditions, there is large production of circulating oxidised LDL particles (oxLDLs). The levels of the latter LDLs have been depicted to be individually prognostic of atherosclerosis (sub-clinical) and coronary heart disease (acute). Studies have established that oxidation of LDL to oxidised LDL particles (oxLDLs) doesn't occur frequently in the circulation. Nevertheless, the same phenomenon is believed to occur in the sub-endothelial space where free radicles derived from cellular metabolism and reactive nitrogen species and reactive oxygen species occur abundantly (Stocker and Keaney Jr 2004). The oxidized LDL in comparison to LDL is caught by the macrophages through scavenger receptors. These macrophage-bound LDL are regarded as lipid laden that is changed to foam cells. Moreover, some of the other properties of oxLDL make it more atherogenic compared to LDL. For instance, this type of oxidized LDL has been found to show chemotaxis potential for monocytes and T cells (Steinberg 2002). It has also been found that oxidized LDL is more immunogenic compared to LDL itself (Matsuura et al. 2008). Reports have depicted that the apo-protein of oxidized LDL, ApoB100 to demonstrate certain epitopes that have a potential to evoke immune responses (Stocker and Keaney Jr 2004). The resulting antibodies produced against the oxidized LDL and associated complexes are also the markers of the cardiovascular complications. Besides oxidation of LDL to oxidized LDL, other immunological phenomenon like inflammation is regarded in the establishment of atherosclerosis (Steinberg 2002). The foam cells are assisted by lymphocytes and smooth muscle cells (vascular) for the formation of plaque (Stocker and Keaney Jr 2004). The lipid fraction as well as the protein part (apoB100) is oxidized during LDL oxidation. The latter phenomenon is encouraged in an environment rich in lipids. It also occurs after taking a diet rich in fats that excites the oxidative stress environments (Campbell et al. 2006). There are reports that suggest that LDL-oxidation is stimulated by obesity and cigarette smoking (Beck et al. 2008). Many compounds including antioxidants have been demonstrated to inhibit this phenomenon (LDL-oxidation) (Campbell et al. 2006; Lapointe et al. 2006). The studies performed in experimental animals are less consistent especially the reports that have shown the role of vitamins and flavonoids (Lapointe et al. 2006). The nutritional style has also been reported to have a role in this phenomenon, for example the Mediterranean-style of food has been observed with lower LDL-oxidation, and however, the lower concentration of carotenoids in serum has depicted a higher oxidation of LDL (Lapointe et al. 2006; Beck et al. 2008). The other molecules that have been reported to inhibit the oxidation of this lipoprotein include melatonin, estrogens and monoamines (Lin et al. 2006).

During the last few decades, the new research insights have demonstrated the role of the diet derived peptides in the LDL-oxidation. In this perspective, some of the peptides have established the inhibitory role of the peptides in the LDL-oxidation. The endomorphins (1 and 2) have already been depicted to inhibit the oxidation of LDL (Lin et al. 2006). It has been found that these peptides possess the activities like scavenging the free radicals and ability to decrease the production of reactive oxygen species (ROS) and induction of lipid peroxidation (Coccia et al. 2001; Fontana et al. 2001). Consequently the milk and other dairy products derived peptides have also been found to have this anti-oxidant property. This includes the peptides from the milk of ovine (Gómez-Ruiz et al. 2008), bovine (Suetsuna et al. 2000), bovine whey and casein (Peng et al. 2008). The hypothesis that BCM-7 derived from the A1 milk of bovines has a role in the LDL-oxidation dates back to Torreilles and Guerin (1995), wherein the researchers found that the LDL is stimulated ex-vivo by the casein hydrolysates (bovine) and tyrosyl residues (Torreilles and Guerin 1995). The tyr-amino acid residues occur frequently in the food. Nevertheless, this study is in contrast to the later findings and that may be because of the experimental status. More research is needed to fully confirm the role of this peptide in the LDL-oxidation.

6.4   Evidences in Favour of A1 Milk Hypothesis and Heart Diseases

6.4.1   Animal Trials

Meeker and Kesten in 1941 demonstrated that when the casein containing diets cholesterol-free) were given to rabbits, these animals developed the atheromatous lesions and hypercholesterolemia (Meeker and Kesten 1941). On the other hand when the same animals were fed with diet containing soy proteins, the results indicated a diminution in both the incidence and degree of atherosclerosis. These findings were later in agreement with other researchers in rabbits and rhesus monkeys that closely resemble human beings (Kritchevsky et al. 1988; Terpstra et al. 1981). Therefore, these findings concluded that the degree of incidence of atherogenicity was due to the varied amino acid composition in casein and soy proteins. The former has a lower ratio of arginine/lysine close to 0.49. However, in the later protein it was higher than 1.18. In this perspective, when the basic amino acid arginine was added to the casein to make the ratio approximately equal, still the results were not meaningful. The results indicated a reduction in atherogenesis by 41% in one group of animals and another group of animals demonstrated a 13% increase in it. Therefore, it was more likely that the increase in the incidence of cardiovascular diseases due to casein was not correlated with the amino acid composition. In the search of alternative hypothesis, the Australian group of researchers performed animal feeding trials in rabbits keeping under consideration the genetic variants of cow milk β-casein. These casein variants were denoted by A1 and A2, the former was regarded as the risk factor due to the release of seven amino acid peptide called BCM-7. The blood vessels in these experimental animals were de-endothelialized before feedings different diets. The diets comprised of 0, 5, 10, and 20% A1 and A2 β-casein (defatted) or the whey protein (hydrolyzed) was kept as control. The diets were added with a total of 20% of the energy in the form of protein and the remaining deficits were compensated with hydrolyzed (partially) whey. Moreover, the diets containing A1 and A2 β-casein or control group animals were added with cholesterol so as to check and compare the harm induced with cholesterol that was putatively correlated to A1 milk or β-casein intake. After 6 weeks of feeding it was found that feeding animals with A1 β-casein without cholesterol demonstrated a remarkably higher LDL-cholesterol, serum cholesterol, HDL-cholesterol and triglycerides compared to animals fed with whey alone. The latter in turn produced higher levels of these molecules in blood compared to animals fed with A2 β-casein. The animals fed with diet having A1 milk β-casein were found to have remarkably higher fatty streaks deposited on the aorta in addition to the thickness of these streaks in the aortic arch compared to animals fed with diet having A2 milk β-casein (Tailford et al. 2003). When cholesterol was added in the diets containing A1 β-casein, results indicated an increase in the thickness of the aortic arch lesion. Therefore, it was assumed that the A1 β-casein from bovines as atherogenic and the cholesterol supplementation further increased the effect in a cumulative manner. Hence it supports the hypothesis in a manner that most western countries take diet rich in A1 milk and dietary cholesterol and there is enough possibility of the increased risk of cardiovascular diseases. On the other hand, the editor of the journal considered this conclusion as premature (Mann and Skeaff 2003). There are certain limitations in the experiment that includes the acute tissue repair that is dominating in this model and this is different from the gradual formation of atherosclerosis progression that follows in man. The fatty streaks developed in the model were from foam cells which in normal atherosclerosis process occur due to macrophages that have engulfed oxidized LDL. In the scratched walls of the blood vessels there can be remarkable oxidative stress that encourages oxidized LDL formation. This animal trial has been observed to be sensitive to the changes in the diet formulation in general (Stocker and Keaney Jr 2004) and the animal used in the trial is specifically sensitive to cholesterol. Consequently, minor alterations in dietary formulation that aren't associated with casein can also have an effect. Keeping in view the limitations of this animal trial it can be put forward that more research with specific atherogenic development models may be designed to reach at a proper conclusion. Also the animals must be fed with A1 and A2 bovine β-casein as well as the commercially synthesized peptides like BCM-7.

6.4.2   Human Studies

The hypothesis that links consumption of A1 milk or derived BCMs with the ischemic heart diseases are mostly based on ecological reports keeping under consideration some environmental agents and occurrence of few non-communicable illnesses. The overall concerns on human associational reports and other supplementary apprehensions on their ambiguities are, therefore, applicable (Morgenstern and Thomas 1993). The associationship between an illness or disorder and its cause through different factors including the behavioral, genetic and environment agent is a very complicated and complex phenomenon. In this regard, explanation of a mono-causal theory for the incidence of a biological complication or disorder is therefore challenging.

The best way to validate a proposed hypothesis is to perform a potential cohort study at the individual level during the course of lifespan, bring together the comprehensive on the diet taken, genetic status of the individual, lifestyle of the person and the environment. This should be followed by the analysis of the complete data and application of the biostatics tools for setting up correlations between these factors and incidence of various biological complications. These comprehensive cohort studies have the limitations in terms of the time taken and the huge amount of the data that has to be collected, that is very cumbersome. In this regard other methods or approaches are taken for performance of these type studies. These alternatives include smaller potential studies and some trials with different conclusions, for instance the risk agents responsible for the illnesses under observation. However, knowledge about these risk agents is difficult to gather as these are either unknown or immeasurable. The other alternative adopted is expressed by case-control reports that correspond with each experimental situation (for example an individual with a disease diagnosed) a measured figure of controls (individuals free from the illness). The other affecting parameters include the age, sex, end point, living area and social status must be as identical as possible between the controls and the cases. However, there must be a large variation in the affecting factor to show an effect that may be notable from the random disparity. However these types of studies must be validated retrospectively the value of supposed affecting agent that is assumed to be responsible for the disease. This may change for different biological complications through different life stages from prenatal age through adulthood, and hence some pieces of potential evidence may be unexplored. The identical type of complications arises when the affecting parameters are assumed but not checked or analyzed. These types of factors that are unknown however a great impact has on the biological complications are regarded as "confounding factors". In this way, this type of study (case-controls) is less valued to establish a cause-effect relationship. The alternative step to check the hypothesis is to perform an ecological study. However, this method has an internal limitation of the collection and analysis of data at the group levels instead of individual level. Although this data is easy to collect and access however implied supposition of uniformity at individual level at large is mismatched with the usage of the observations of the study to establish a cause-effect relationship.

This may be the preliminary step to gather the information or evidences for establishment of the hypothesis. The progressions that lead to the many potential biological complications like cardiovascular diseases (CVD) including stroke and ischaemic heart disease are many causes and are created over extensive period before these illnesses are diagnosed, more notably, after 50–60 years of age. CVD is regarded as a lifestyle illness. It occurs in most of the countries and especially in the countries with high living standards, the incidence of this illness is found very high and is observed a major cause of mortality (WHO 2003). This disorder has already been comprehensively studied from decades, many risk agents have been established fully that develop over long period of time and even measured from the childhood. These risk agents include the high lipoprotein (LDL-cholesterol) and low HDL-cholesterol, high blood pressure, increased expression of inflammatory markers and homocysteine, decrease in physical activity, smoking and obesity.

Similarly, the enhanced incidence of obesity in the young individuals is linked with the increased prevalence of the type 2 diabetes. The latter individuals have been found to be susceptible for incidence of CVD. The recent research links type 2 diabetes and CVD with various immunological responses including inflammatory mechanism and the metabolic dysfunctions in the organs. The overlying mechanisms for these dysfunctions hasn't been completely comprehended or explored, nevertheless, these illnesses are linked with the incidence of chronic inflammation as one of the potential induction factors. The possible role of the nutrition of an individual in regulating the incidence, development and prevalence of the inflammatory responses has already been published (Nicklas et al. 2005).

The milk is the sole food for infants and a part of diet for adults also. The cow milk has also been a part of diet for infants directly in the liquid form or in the form of infant formulas. In this perspective some ecological correlations have associated consumption of A1 milk with the incidence of cardiovascular diseases. These reports have estimated the consumption of A1 β-casein from the different breeds of cows with probable β-casein variants in a particular region or country and applied statistical analysis tools for the establishment of the correlation. The first report was collected in 1981 by Seely and coworkers for the establishment of the associationship between milk protein consumption and mortality with CVD (Seely 1981). This was followed by the McLachlan in 2001, who hypothesized the same correlation between consumption of A1 β-casein and ischaemic heart disease mortality as noted 5 or 10 years later in 16 selected countries around 1980. The investigators found that when the individuals consumed low-fat diet (considered rich in disease-risk factor), the unexpected results were found i.e., a reduction in mortality couldn't be observed always. The analysis for the mortality (with IHD) was performed according to the WHO report published in 1995 on CVD for elderly. This was followed by a 5-year lag period between consumption of cow milk (risk factor) and mortality rate. The authors reported a weak correlation ($r^2 = 0.26$) between consumption of cow milk and incidence of IHD when all countries were included in the statistical analysis. However, to a surprise the correlation was found to be very strong ($r^2 = 0.71$) between consumption of cow milk and incidence of IHD when only those countries were included in the statistical analysis that contained cow milk with specifically A1 variant of β-casein. The correlation was developed more strong ($r^2 = 0.84$) when the cheese was excluded in the analysis. The latter dairy product was excluded in the study due to the varied consumption of this product in different regions or countries. For example the French and the Swiss consumers take more cheese compared to English that consumed substantially less. Moreover, in the same ecological report, the authors assumed that there could be a close associationship between smoking, the intake of A1 β-casein and the mortality due to cancer and CVD (McLachlan 2001). These observations were however questioned later on the grounds that the selected countries varied in incomes and capacities to deliver coronary care. Laugesen and Elliott in 2003a, b also established an associationship between consumption of A1 milk and development of IHD in twenty countries. This study excluded the consumption of cheese and butter. This study also couldn't consider consumption of tobacco and correlated it with incidence of CVD. Therefore, these points lead to the generalizations that ecological studies are unable to regulate all the confounding agents. The result of this study and the mortality rates due to IHD due to the consumption of A1 milk protein was predictable. Japan was found to have lower mortality rate due to IHD and the consumption of A1 milk protein and saturated fat was correspondingly lower in this country. The mortality rates due to IHD in the subsequent lowermost graded countries also corresponded well with the A1 milk protein consumption, excluding Switzerland. Although the IHD rates in the France and the Island were lower in spite of higher consumption of milk, nevertheless the milk consumed contained lower fractions of A1 β-casein. Quite expectedly, the individuals living in the countries with highest IHD rates in the study including United Kingdom, Finland, Ireland and New Zealand corresponded well with the milk intake with high A1 β-casein. The supply of milk protein and the type of β-casein in the milk of different cows correlated strongly with the incidence of IHD in comparison with the dietary fat, smoking and alcohol consumption. The statistical (multivariate) analysis including all the data for 20 years depicted a strong correlation incidence of IHD and consumption of milk with high A1 β-casein ($r = 0.76$; $P < 0.0001$) (Laugesen and Elliott 2003a, b). While utilizing the data from the year 1970, Hill and coworkers in 2002 worked for the correlation between consumption of A1 β-casein and CVD prevalence. Nevertheless, the correlation was found to be weak when data was taken from 1990. The researchers therefore concluded that the comparative quantity of A1 β-casein was deceptively increasing in cow milk during this time span in these countries, the intake of this type of cow milk protein had no influence on the mortality rate due to IHD (Hill et al. 2002). Contrariwise, to the aforementioned reports, the more new cohort reports have established a protective role of cow milk intake for stroke and heart diseases (Elwood et al. 2004a, b, 2005). In the last decade, two human interpolation studies have been performed to confirm the hypothesis of A1 milk intake and incidence of CVD. In one of the blind crossover study, 15 individuals were selected for the study and given A1 β-casein at a dose of 25 g/day, while the other group of individuals was given the same dose of A2 β-casein for a period of 12 weeks. The results couldn't find any significant difference in any of the functional and endothelial tests between the two groups of human subjects (Chin-Dusting et al. 2006). Similarly, Venn and coworkers in 2006 tried to validate the effect consumption of A1 and A2 cow milk β-casein on serum cholesterol levels in humans. The study involved 62 human subjects for a period of about 31 days. The results indicated non-significant changes in serum cholesterol levels on intake of A1 or A2 β-casein (Venn et al. 2006). These types of studies are considered to provide strong piece of evidence, and couldn't provide strong proofs for the already established hypothesis. Therefore, it may be concluded that more research with advanced technology is needed to fully confirm the laid hypothesis.

6.5 Possible Mechanism for Decreasing the Risk of Cardiovascular Disease

One of the most likely justification that cow milk β-casein could be atherogenic are the in vitro findings that demonstrated the oxidation of LDL-cholesterol to oxidized LDL-cholesterol (Torreilles and Guerin 1995). This biochemical finding was later validated through immunological perspective, where antibodies were assessed in infants against the oxidized LDL-cholesterol. The levels of these latter molecules were remarkably linked between the mother and their offsprings at the time of birth ($r = 0.79$; $P < 0.001$). Nevertheless, at the age of 3 months, the level of these antibodies in the serum was dependent on breast-feeding or bottle-feeding. The infants who took infant formulas composed of cow milk depicted remarkably higher antibody levels in comparison to those who got breast-feeding. Therefore, these findings paved a way for explanation of the promotion of oxidation of LDL-cholesterol, development of raised antibodies against the latter and an enhancement in the risk of CVD. More importantly, the production of oxidized LDL-cholesterol that is catalyzed by tyrosyl radicals (amino terminal) may be the chief process in the establishment of atherosclerosis (McLachlan 2001). In this perspective, the BCM-7 is released from the A1 "like" variants of cow milk β-casein and possesses an amino terminal with tyrosyl end amino acid residue. This β-casein derived peptide has been shown to demonstrate the catalytic property that leads to oxidation of LDL-cholesterol to oxidized LDL-cholesterol (Torreilles and Guerin 1995). The other possible explanation for the cow milk as risk factors for heart diseases is the stimulation of platelet aggregation (McLachlan 2001). Moreover, there are other factors that impart some effects on the pulmonary system are the peptides derived from cow milk β-casein (Fiat et al. 1989; Meisel and FitzGerald 2000). The dominating factor for the implication of heart diseases by the β-casein derived peptides is the oxidation of LDL-cholesterol, the other factors are supplementary.

6.6 Critique of the A1 Milk Hypothesis and Heart Disease

- The between-country relationships were regarded as unreliable as these couldn't establish cause-effect relationships.
- For example, the death rate due to CHD in Jersey is about the same as in Australia, inspite of higher consumption of A1 β-casein (41% A1 in whole) in the later compared to the former.
- When correlation coefficients were recalculated for various foods by Crawford and coworkers in 2003, giving time lag periods and higher set of countries, the association between milk protein consumption and mortality due to CHD couldn't lessen to zero, probably due to latest alterations in the coronary mortality rates in the countries.

- A time lag period of 5 years was given by the authors McLachlan in 2001 and Laugesen and Elliott in 2003a, b between consumption of A1 β-casein and mortality due to CHD mortality, however, this period of time is considered as a short duration (McLachlan 2001; Laugesen and Elliott 2003a, b).
- Laugesen and Elliott in 2003a, b couldn't build a correlation between consumption of tobacco and CHD mortality. The tobacco consumption or smoking is considered the well-established risk factors, therefore, difficult to rely on these correlations (Laugesen and Elliott 2003a, b).
- While describing the observations of Laugesen and Elliott in 2003a, b, few concerns were explained as below.

  - Many of the countries didn't follow the average regression line.
  - The two countries Austria and France were found to have same milk protein β-casein intake (0.93 g/day) but mortalities due to CHD was found to be 88 and 33 per 100,000, respectively.
  - Many countries including Australia, Sweden, Iceland, Germany, Canada, Israel and the island of Jersey were found to have mortalities due to CHD between 70 and 80 per 100,000. However, the A1 β-casein intake was found to the highest in the world i.e., 2.8 g/day in Sweden and the lowest i.e., 0.3 g/day in Jersey (Laugesen and Elliott 2003a, b).

- The mortalities due to CHD in numerous countries have been analyzed closely, in some of the countries like USA, Finland and Australia, the rates have come down, while, in some countries of eastern Europe have increased. In Switzerland, there is authentic data on milk intake and on human illnesses. The milk intake has been constant at about 25 g/individual/day, the mortality due to CHD has gone down gradually from 160 per lac in 1980 to 95 per lac per year in the middle of 1900s (Hill et al. 2002).
- The data taken from WHO regarding the mortalities due to CHD and intake of dietary components like milk protein, cheese, alcohol, coffee and meat from 47 countries. These associationships were calculated for these countries for every second year from 1969 to 1995. These associationships were calculated for the same year data for consecutive time lag up to 30 year. The results of these computations were amazing and the associations with milk protein intake were close to 0.7 till 1983. From the following year this figure has come down to zero. The researchers speculate that in the most countries the fraction of A1 β-casein in the whole milk is about 30–55% of total β-casein. Therefore, a correlation may be set up for CHD mortalities if the intake of A1 β-casein is a causative agent. The various lag periods given made small difference in the established correlations and the other dietary elements checked demonstrated positive but lower correlations. These correlations decreased since 1983 and currently are below zero (Crawford et al. 2003).
- Regarding the animal experimentation, only one animal trial has been performed so far regarding A1 and A2 milk protein β-casein (Tailford et al. 2003). In this experiment, several lacunas were observed.

- The animal trial with rabbits was performed for short duration (only for 6 weeks).
- The number of animals in each group were lesser i.e., six per group.
- The animals (rabbits) have a normal diet in the form of leaves. But in the experimental trial artificial diet was given to these animals composed of animal proteins (20%).
- The rabbits used in the experiment for the incidence of atherosclerosis are not the correct model for this complication, in human this develops gradually over years in humans in the form of patches of fatty streaks.
- Although the raised serum cholesterol levels were observed in rabbits on feeding experimental diet. However, this can't be applied to humans that the same serum cholesterols levels would increase on administration of the same diet.
- The assessment of the fatty streaks in the aorta wasn't prepared blind for the experimental (diet) group and the variation in the regions between A1 and A2 groups weren't significant in the aortas of the animals that were administered with the cholesterol in the diets, or in the carotid arteries.
- Mann and Skeaff in the year 2003 in an editorial explained that there were remarkable limitations in concluding these observations to clinical effects in humans (Mann and Skeaff 2003).

- The collection of many (seven) reports composed by Ness and coworkers in 2002 examined the link between milk intake and cerebrovascular and coronary mortalities. In none of the studies the milk intake are correlated with total or cardiovascular mortality (Ness et al. 2002).
- The WHO report 2003 on the chronic diseases and human health and nutrition listed 23 factors (dietary) that are linked or not linked to cardiovascular diseases (WHO/FAO Consultation 2003). Milk was not included in the list. Quite evidently, the developed countries bovine milk is mostly A1, therefore, bovine milk isn't generally regarded as a risk factor responsible for development of cardiovascular disease.

References

Allison AJ, Clarke AJ (2006) Further research for consideration in "the A2 milk case". Eur J Clin Nutr 60(7):921–924

Beck J, Ferrucci L, Sun K et al (2008) Circulating oxidized low-density lipoproteins are associated with overweight, obesity, and low serum carotenoids in older community-dwelling women. Nutrition 24(10):964–968

Campbell CG, Brown BD, Dufner D et al (2006) Effects of soy or milk protein during a high-fat feeding challenge on oxidative stress, inflammation, and lipids in healthy men. Lipids 41(3):257–265

Chin-Dusting J, Shennan J, Jones E et al (2006) Effect of dietary supplementation with [beta]-casein A1 or A2 on markers of disease development in individuals at high risk of cardiovascular disease. Br J Nutr 95(1):136–144

Coccia R, Foppoli C, Blarzino C et al (2001) Interaction of enkephalin derivatives with reactive oxygen species. Biochim Biophys Acta 1525:43–49

Crawford RA, Boland MJ, Norris CS et al (2003) Milk containing beta casein with proline at position 67 does not aggravate neurological disorders. PCT/WO 02/19832/A1

Elwood PC, Pickering JE, Fehily AM et al (2004a) Milk drinking, ischaemic heart disease and ischaemic stroke I. Evidence from the Caerphilly cohort. Eur J Clin Nutr 58(5):711–717

Elwood PC, Pickering JE, Hughes J et al (2004b) Milk drinking, ischaemic heart disease and ischaemic stroke II. Evidence from cohort studies. Eur J Clin Nutr 58(5):718–724

Elwood PC, Strain JJ, Robson PJ et al (2005) Milk consumption, stroke, and heart attack risk: evidence from the Caerphilly cohort of older men. J Epidemiol Community Health 59(6):502–505

Fiat AM, Levy-Tolediano S, Caen JP et al (1989) Biologically active peptides of casein and lactoferrin implicated in platelet function. J Dairy Res 56:351–355

FitzGerald RJ, Meisel H (2000) Milk protein-derived peptide inhibitors of angiotensin- I-converting enzyme. Br J Nutr 84(1):33–37

Foltz M, Meynen EE, Bianco V et al (2007) Angiotensin converting enzyme inhibitory peptides from a lactotripeptide-enriched milk beverage are absorbed intact into the circulation. J Nutr 137(4):953–958

Fontana M, Mosca L, Rosei MA (2001) Interaction of enkephalins with oxyradicals. Biochem Pharmacol 61(10):1253–1257

Gómez-Ruiz J, López-Expósito I, Pihlanto A et al (2008) Antioxidant activity of ovine casein hydrolysates: identification of active peptides by HPLC–MS/MS. Eur Food Res Technol 227(4):1061–1067

Hartmann R, Meisel H (2007) Food-derived peptides with biological activity: from research to food applications. Curr Opin Biotechnol 18(2):163–169

Hill JP, Crawford RA, Boland MJ (2002) Milk and consumer health: a review of the evidence for a relationship between the consumption of beta-casein A1 with heart disease and insulin-dependent diabetes mellitus. Proc NZ Soc Anim Prod 62:111–114

Hong F, Ming L, Yi S et al (2008) The antihypertensive effect of peptides: a novel alternative to drugs? Peptides 29(6):1062–1071

Korhonen H, Pihlanto A (2006) Bioactive peptides: production and functionality. Int Dairy J 16(9):945–960

Kritchevsky D, Tepper SA, Story JA (1988) Influence of soy protein and casein on atherosclerosis in rabbits. Fed Proc 15(3):163–169

Lapointe A, Couillard C, Lemieux S (2006) Effects of dietary factors on oxidation of low-density lipoprotein particles. J Nutr Biochem 17(10):645–658

Laugesen M, Elliott R (2003a) Ischaemic heart disease, type 1 diabetes, and cow milk A1 β-casein. N Z Med J 116(1168):U295

Laugesen M, Elliott R (2003b) The influence of consumption of A1 β-casein on heart disease and type 1 diabetes–the authors reply. Z Med J 116(1170):U367

Lin X, Xue LY, Wang R, Zhao QY, Chen Q (2006) Protective effects of endomorphins, endogenous opioid peptides in the brain, on human low density lipoprotein oxidation. FEBS J 273(6):1275–1284

Mann J, Skeaff M (2003) β-Casein variants and atherosclerosis—claims are premature. Atherosclerosis 170(1):11–12

Matsuura E, Hughes GRV, Khamashta MA (2008) Oxidation of LDL and its clinical implication. Autoimmun Rev 7(7):558–566

McLachlan CNS (2001) β-Casein A1, ischaemic heart disease mortality, and other illnesses. Med Hypotheses 56(2):262–272

Meeker DR, Kesten HD (1941) Effect of high protein diets on experimental atherosclerosis of rabbits. Arch Pathol 31:147–162

Meisel H, FitzGerald RJ (2000) Opioid peptides encrypted in intact milk protein sequences. Br J Nutr 84(1):27–31

Moller NP, Scholz-Ahrens KE, Roos N et al (2008) Bioactive peptides and proteins from foods: indication for health effects. Eur J Nutr 47(4):171–182

Morgenstern H, Thomas D (1993) Principles of study design in environmental epidemiology. Environ Health Perspect 101(4):23–38

Ness AR, Smith DG, Hart C (2002) Milk, coronary heart disease and mortality. J Epidemiol Community Health 55:379–383

Nicklas BJ, You T, Pahor M (2005) Behavioral treatments for chronic systemic inflammation: effects of dietary weight loss and exercise training. Can Med Assoc J 172(9):1199–1209

Peng X, Xiong YL, Kong B (2008) Antioxidant activity of peptide fractions from whey protein hydrolysates as measured by electron spin resonance. Food Chem 113(1):196–201

Saito T (2008) Antihypertensive peptides derived from bovine casein and whey proteins. Adv Exp Med Biol 606:295–317

Seely S (1981) Diet and coronary disease: a survey of mortality rates and food consumption statistics of 24 countries. Med Hypotheses 7(7):907–918

Steinberg D (1997) Low density lipoprotein oxidation and its pathobiological significance. J Biol Chem 272(34):20963–20966

Steinberg D (2002) Atherogenesis in perspective: hypercholesterolemia and inflammation as partners in crime. Nat Med 8(11):1211–1217

Stocker R, Keaney JF Jr (2004) Role of oxidative modifications in atherosclerosis. Physiol Rev 84(10):1381–1478

Suetsuna K, Ukeda H, Ochi H (2000) Isolation and characterization of free radical scavenging activities peptides derived from casein. J Nutr Biochem 11(3):128–131

Tailford KA, Berry CL, Thomas AC et al (2003) A casein variant in cow's milk is atherogenic. Atherosclerosis 170(1):13–19

Terpstra AHM, Harkes L, Van Der Veen FH (1981) The effect of different proportions of casein in semi purified diets on the concentration of serum cholesterol and the lipoprotein composition in rabbits. Lipids 16:114–119

Torreilles J, Guerin MC (1995) Casein-derived peptides can promote human LDL oxidation by a peroxidase-dependent and metal-independent process. CR Seances Soc Biol Fil 189(5):933–942

Venn BJ, Skeaff CM, Brown R et al (2006) A comparison of the effects of A1 and A2 β-casein protein variants on blood cholesterol concentrations in New Zealand adults. Atherosclerosis 188(1):175–178

WHO (2003) Diet, nutrition and the prevention of chronic diseases. Report of the WHO/FAO Joint Expert Consultation WHO Technical reports 916: http://whqlibdoc.who.int/trs/WHO_TRS_916.pdf

Chapter 7
A1 Milk and Neurological Diseases

Abstract Many ecological correlations have correlated consumption of A1 milk with increased risk for incidence of neurological manifestations like autism, schizophrenia and SIDS. These correlations are supported by human case-control studies. It is assumed that BCM-7 released from A1 milk is actually the hypothetical risk element. BCMs are proposed to produce in the gut, transport to blood, passage across the blood-brain barrier, act on the brain cells and subsequently induce the neurological manifestations. Studies have established that the antibody titers (IgG) against BCM-7 are present in major of the patients (90%) and (86%) suffering from schizophrenia and autism respectively. Additional reports have demonstrated remarkably increased levels of BCM-7 in the blood and cerebral spinal fluid in these patients compared to normal subjects. However, few American Nutritionists, EFSA and Truswell refute all these correlations and mechanisms.

7.1 Autism

This disorder is one among the Autistic spectrum disorders (ASDs), the latter further comprises of atypical autism and few allied complications in addition to Asperger's disorder (Cass et al. 2008). Autism is regarded as an intricate neurodevelopmental condition in which patients demonstrate reduced give-and-take communicational behaviours in addition to limited, monotonous or stereotyped performances (Barbaresi et al. 2006). A heterogeneous population of youngsters suffering from autism is usually observed. The issue of the neurological condition autism and allied complications stands potentially important and intriguing due to enthused community apprehensions regarding the actual enhancement in the autistic cases and associated disorders. The mechanism of the induction of ASDs is quite indistinguishable; moreover, the psychological or therapeutic practices yield unsatisfactory consequences. This is due to this probable reason also that much consideration is given to dietary factors, additional environmental agents and complementary treatments (Barbaresi et al. 2006). In this perspective, one of the concepts known as "leaky gut" conception has gained enough consideration. This perception has further been extended to another proposed reason that is the perfor-

© Springer Nature Singapore Pte Ltd. 2020
M. R. Ul Haq, *β-Casomorphins*, https://doi.org/10.1007/978-981-15-3457-7_7

mance of vaccination. For instance, there has been enough debate going on the potential role of measles-mumps-rubella (MMR) vaccine and induction of autism. According to this hypothesis, the latter vaccination enhances the permeability of gut causing "leaky gut". However, there isn't enough epidemiological or ecological data to build a correlation between induction of autism and MMR vaccination. The concept of "leaky gut" is further strengthened by the studies of Cade and coworkers who observed peaks in the urine samples with increased number of peptides in such neurological patients including schizophrenia and autism (Cade et al. 2000). Nevertheless, the fractions that contain these peaks haven't been characterized comprehensively. Consequently, the nutritional interference that seemingly led to ameliorations was free from dairy and wheat linked casein and gluten respectively. Moreover, this concept is still dominating and has been the main foundation of the diet formulation for ASD subjects. This concept is further supported by the studies of Reichelt and Knivsberg who proposed a model known as "autism model" that proposes that food derived peptides (exorphins) and serotonin uptake modulators are the key players in the establishment of autism (Reichelt and Knivsberg 2003). They further added that there is some genetic base to strengthen the model in terms of the gene deficiencies that expresses at least two or more enzymes (peptidases) and/or some other proteins that regulate these enzymes. On the other hand this proposed model has been a subject of controversy by many other investigators. The subsequent reports with application of more sophisticated bioanalytical techniques including LC/MS couldn't detect these peptides. The other investigators also couldn't confirm the presence of these opioid peptides in the urine of autistic patients. They also couldn't develop a correlation between development of autism and deficiency of the brush border enzymes like dipeptidyl peptidase IV (Hunter et al. 2003). The other research group exploited mass spectroscopy (MALDI-TOF/MS) for analysis of such peptides (opioid). However, the investigators couldn't find any evidence of such peptides in the urine of autistic patients. They finally concluded that these opioid peptides can't be regarded as marker (biomedical) for such diseases. These peptides may not be engaged to analyse the response to diets containing casein or gluten (Cass et al. 2008). The other researchers in 2006 thoroughly studied the diet formulations (gluten and casein-free) and concluded that there aren't enough evidences to formulate recommendations about these diet elements. Moreover, they stated that the casein and gluten-free diets should be analyzed first and the consequential parameters must comprise of evaluation of non-verbal cognizance (Christison and Ivany 2006). In 2007, Dettmer and coworkers while working on food derived opioid peptides like BCMs, gliadinomorphin, deltorphin-1 and deltorphin-2 couldn't find these peptides in the urine of autistic children (Dettmer et al. 2007). Subsequently, Millward and coworkers published a review in this research area. They concluded that the application of complementary and alternative therapies (CAM) for autistic patients with casein and gluten-free diet formulations couldn't provide enough encouraging results (Millward et al. 2008). This scientific literature in this research area isn't in a position to formulate recommendations regarding establishment of autism and intake of gluten and/or casein-free diets.

7.2   Exorphins and the Central Nervous System (CNS)

7.2.1   Distribution of Receptors for Opioids in the CNS

Receptors for endorphins or exorphins are extensively located in the brain, spinal cord, autonomous nerves and peripheral sensory nerves (Wittert et al. 1996; IUPHAR 2008; Peckys and Landwehrmeyer 1999). These receptors through their corresponding ligands (endorphins and exorphins) have numerous and rather varied actions and manifestations. These effects include sedative actions, analgesic actions, euphoria, eating and appetite behaviours, dysphoria, depression in respiration, cough reflexes, vomiting, nausea and pupillary constriction. The mu-receptors (related to morphine agonists) are reported to interact with the milk derived exorphins like BCMs (Brantl et al. 1981, 1982; Koch et al. 1985). These μ-receptors for BCMs are distributed in the caudate putamen, thalamus, nucleus accumbens neocortex, interpeduncular complex, amygdala, and inferior and superior colliculi (IUPHAR 2008). The receptors (μ) are also expressed in the dorsal horn (the superficial layers) of the spinal cord. Moreover, a medium density of these receptors is also expressed in raphe nuclei and periaqueductal gray. These areas in the brain have well-developed role in the induction of pain and analgesic effects. Morphine and other opioid compounds including fentanyl are classified in the ligands that bind μ-receptors. Besides these effects, these demonstrate role in other processes including respiration, cardiovascular functions, transit of intestine, behaviours, mood changes, heat regulation, secretion of hormones and some immune effects.

7.2.2   Action of Exorphins on CNS

There are many animal reports that show the role of bovine milk derived BCMs in the CNS. The peptides were administered to these experimental animals parenterally, except for the animal report where the effect of soymorphins peptides was observed by Ohinata and coworkers (Ohinata et al. 2007). It was Brantl and coworkers for the first time in 1981 and parallely Grecksch and coworkers in the same year who reported the analgesic effect of BCMs when these peptides were injected into the intracerebro-ventricules. The work was more enthusiastic because of the reason that the study didn't include single but four different BCMs ranging from four to eight amino acid residues marked as BCM-4, BCM-5, BCM-7 and BCM-8. The results were promising as the trials revealed analgesic effect of these peptides. Additionally the effect was entirely reversed by the μ-opioid receptor antagonist, naloxone. Moreover, the relative responses revealed that although BCM-7 demonstrated slowest commencement of action, but the same response occurred for longer time duration of 90 min. The investigators highlighted based on the long time duration that lasted even more than endorphins they have assessed. Paroli in 1988 observed sedative and analgesic effects of these peptides on CNS when administered directly in blood (Paroli 1988). Moreover, initial neonatal interaction of experimental animals (rats) to the opioid peptide (morphiceptin) induced hyper analgesic condition in later life (4.5 months) (Zadina et al. 1997). These observations therefore infer that the correlating this animal data to humans need more well-designed studies, because of the wide variations in developmental stages of neonates (humans). Based on these preliminary animal experiments, many researchers validated the role of these bovine milk derived BCMs in the central nervous system of experimental animals when administered parenterally. The full details are shown in Table 7.1.

7.3 A1 Cow Milk Consumption and Neurological Disorders

Many correlations that have associated consumption of cow milk β-casein with neurological diseases including autism, schizophrenia, SIDS and post-partum depression (Sun et al. 1999, 2003; Sun and Cade 2003). These complications are very significant for considerations as only little treatment choices are available and remarkably very annoying for the family members. There are some studies available that have tried hard to demonstrate the relation, but this maiden data is challenging. The researchers at the Florida University postulated that the BCM-7 crosses the blood-brain barrier and acts on the brain cells and hence induces the symptoms linked with autism, schizophrenia and SIDS (Sun et al. 1999, 2003; Sun and Cade 2003). Studies have established that the antibody titers (IgG) against BCM-7 are present in major of the patients (90%) and 86% suffering from schizophrenia and autism respectively. Additional reports have demonstrated remarkably increased levels of this peptide (BCM-7) in blood and cerebral spinal fluid in these patients compared to normal subjects. The increase in the levels of BCM-7 and in turn enhanced antibody titers may be due to the deficiency or lacks of enzymes that have specificity for these proline rich peptides like dipeptidyl peptidase-4. Alternatively, there may be increased immature gut or increased gut permeability to allow these peptides to cross the gut barrier and reach the circulation. It is also possible that cumulative action of the two mechanisms may work together to correlate BCM-7 with neurological conditions. In order to fully rely on the hypothesis that regards this opioid peptide is associated with the neurological symptoms. It is significant to characterize, confirm and validate the findings that this peptide crosses the blood brain barrier. To validate the findings that this opioid peptide really crosses the barrier between blood and brain in a carrier-dependent manner (Banks and Kastin 1987; Pasi et al. 1993). The other possibility is that it is not essential for the whole opioid peptide to cross this barrier to create these complications, instead only a subgroup of the opioid peptide is essential to cross.

Table 7.1 Effect of bovine milk derived BCMs on central nervous system in different animals when administrated through different routes

Peptides	Effect	Literature
BCMs (4–8)	When these peptides were administered in rats through intracerebro-ventricular route, they showed analgesic activity, the effect was antagonized on naloxone administration	Brantl et al. (1981)
BCM-5	When these peptides were administered in rats through intracerebro-ventricular route, they observed analgesic activity that was dose-dependent and the effect was antagonized by naltrexone	Grecksch et al. (1981)
BCMs (4–7)	When these peptides were administered in young chickens, through intracerebro-ventricular route, they decreased the separation prompted distress vocalizations (DVs). Moreover, the results indicated that BCM-5 to be more active than other two analyzed peptides. However, BCM-7 demonstrated the prolonged effect. These effects were partly reversed on naloxone administration	Panksepp et al. (1984)
BCM-5	When this peptide was administered through intracerebro-ventricular and intravenous routes, the results demonstrated the analgesic effect. However, BCM-5 was found to be 20 times less active compared to morphine when given through intracerebro-ventricular route	Matthies et al. (1984)
BCM-7	After intraperitoneal administration, this peptide didn't show any effect on waking. At a concentration of 100 mg/kg, the symbols of sleep altered. No symptoms of any respiratory depression occurred. The effect was antagonized on naloxone pretreatment	Taira et al. (1990)
BCM-7	After intraperitoneal administration of this peptide in rats, a quickening effect in learning was demonstrated for procurement of food habit in a T-maze. Results also indicated a late stimulus in the delayed learning of active avoidance reply that includes the usage of a hurting underpinning provocation	Maklakova et al. (1995)
BCM-5	When this peptide was given through intracerebro-ventricular route, amnesia in experimental mice was observed that was dose-dependent When this peptide was given intraperitoneally at a dose of 0.1–20 mg/kg. Non-significant changes were observed in impulsive alternation behavior On the other hand, other studies showed that injection of this peptide intraperitoneally at a low dose of 1 mg/kg ameliorates disruption of learning and remembrance via central intermediation through mu-opioid receptor pathway. BCM-7 was observed to encourage neurite extension on neuroblastoma cells	Sakaguchi et al. (2003a, b, 2006)
BCM-7	When given to rat pups intraperitoneally demonstrated postnatal learning. This peptide has a considerable and extended activity related to adaption for mammals (young) with respect to environment	Dubynin et al. (2008)
BCM7	When given to newborn rats intraperitoneally resulted in synthesis of DNA (ex-vivo activation) in myocardium and epithelium of ectoderm and endoderm of rats (newborn)	Maslennikova et al. (2008)

7.3.1 *Animal Studies*

The animal trials were carried out in rats and administered with different doses of BCM-7 opioid peptide. The presence of this peptide in different regions of the brain was assessed in terms of immuno-reactivity (FOS like) (Sun et al. 1999). The effect was expressed as the regions affected when there was an increase in the number of stained cells compared to the regions of brain taken from rats infused with normal saline. The results indicated increase in the regions of brain affected by BCM-7. The temporal and occipital cortex (speech center) was observed to be strongly affected. Moreover, it is quite evident that the patients with autism and schizophrenia have abnormal patterns of speech. Additionally, there is a disruption in the levels of dopamine, serotonin and γ-aminobutyric acid (GABA), the disturbance in the latter leads to diminished or lack of social interactions that are representative of the neurological conditions. This is quite clear that this peptide is released in gastrointestinal tract from A1 β-casein. However, other peptides resembling this peptide are released from wheat gluten and named as gliadinomorphins. The investigators postulated that a remedy can be acquired from these neurological conditions while taking a diet free from gluten and casein (Sun et al. 1999, 2003; Sun and Cade 2003). The same research group in 1999 administered intraperitoneally BCM-7 to rats to check the changes in behaviour or demonstrate analgesic effect (Sun and Cade 2003). The results indicated significant changes after time intervals (2 h). With first minute, the animals showed restlessness and enhanced respiratory. The animals were observed to be lethargic and demonstrated decreased social interaction after 7 min. There were non-significant changes observed in rats infused with normal saline. Moreover, the rats co-administered with BCM-7 and naloxone (BCM-7 antagonist) also exhibited non-significant changes. This is due to the fact that naloxone blocks the μ-opioid receptor pathway exploited otherwise by opioid peptide BCM-7. These observations therefore support the role of this opioid peptide BCM-7 in the pathogenesis of neurological complications like autism and schizophrenia. The same investigators in an attempt tried to explain the associationship between BCM-7 and SIDS (Sun et al. 2003). BCMs in general and BCM-5 and BCM-7 in particular have been found to be strong opioid agonists. More importantly, BCM-5 is found to be around ten times more potent compared to morphine and the other peptide (BCM-7) almost equally potent. However, there is insufficient data that clearly regards BCM-7 as a hypothetical risk factor for SIDS. The discussion on whether A1 β-casein from bovine milk releases BCM-7 during gastrointestinal digestion and then the latter correlates with SIDS is both potentially important and intriguing. Some animal trials have depicted a depression in respiration resembling that of SIDS when BCMs are directly administered in intra-cerebroventricular cavity. On the other hand, there are other reports that have correlated consumption of BCMs and SIDS positively. Studies have also shown high levels of BCMs in the cerebrospinal fluid of respiratory distress infants compared to normal ones. On one hand insufficient findings can't conclude a strong correlation between BCM consumption and SIDS. But on the other hand the data depicting intake of bovine milk, respiratory distress and changes in behaviour can't be overlooked. Therefore, to establish a strong correlation between the BCM intake and SIDS, the presence of these peptides in the cerebrospinal fluid, plasma or serum should be estimated with advanced techniques like mass spectroscopy, ELISA and analytical HPLC methods. Moreover, well-designed human trials should be performed involving neonates with disturbances in respiration and healthy neonates for perfect assessment. There are also few ecological studies that link A1 bovine milk consumption and neurological manifestation like schizophrenia and autism. Reports have also estimated higher levels of this peptide (BCM-7) in the blood (Lindstrome et al. 1984; Reichelt et al. 1990) and urine (Cade et al. 2000; Reichelt et al. 1991) of subjects suffering from neurological complications including autism, schizophrenia and postpartum psychosis. To date the possible mechanism formulated claims the release of BCM-7 during gastrointestinal digestion from A1 bovine β-casein, transport of this peptide through gut, entry into the circulation, crossing the BBB and finally expressions of neurological disorders (Sun et al. 2003). The hypothetical link between A1 milk consumption and schizophrenia and ASD doesn't tell about cause of the neurological conditions but to the exacerbation of signs linked with these neurological complications. More specifically, it may be concluded that this peptide may exaggerate indications connected with these complications. There are also a number of laboratories located in USA and Europe that assess BCM-7 levels in urine with a purpose of nutritional intermediation for the management of ASD patients. In this perspective, Cade and coworkers in 2000 revealed that a diet free from gluten and casein was complemented by amelioration in 81% of autistic patients within 3 months (Cade et al. 2000).

7.3.2   BCMs and Sudden Infant Death

The depression in respiration with opioids is a well-established fact. In medical terms it may be put forward that depression in respiration decreases the usage of analgesia (opioid). The principle effort for respiration is produced in the brain stem. This is controlled by elements which comprises of conscious feedbacks from the central (brainstem), cortex, and peripheral receptors (chemo) which sense alterations in the biochemical elements of the blood (Pattinson 2008). There are number of ways through which opioid weaken the respiration and includes various sites (neuronal) on which they may show mechanism of action. Also some differences are manifested due to the diversity in the opioids and that may exist in a specific group also (e.g. μ-receptor binding molecules). More generally there are some correlations between sleep disturbances and apnea and use of opioids (Wang and Teichtahl 2007). The receptors for opioids and sleep regulation are positioned in the same regions of the brain (nuclei) and these peptides (opioid) are proposed to have a role in the stimulation and maintenance of the sleep state. Whenever there is a change in stimulation, maintenance and removal of opioids, irregular sleep patterns have been stated. The reduction in REM (rapid eye movement) sleep has been observed during opioid induction and maintenance. In addition to this central sleep apnea (CSA) has been documented with use (chronic) of opioids.

The mechanism of SIDS induction is complicated and multifactorial. It is the source of mortality of newborns from first month to first year of their age (Brooks 1982). Sun and coworkers in 2003 stated that the common factor for all the children that suffer with SIDS is the consumption of milk. The latter is the complete diet of the infants, when digested in their gastrointestinal tract and release BCMs. Since the gut of the infants is immature therefore possess leaky gut that provides enough chances for these peptides to cross the gut reach blood and finally cross BBB. In the brain of those infants that have irregular respiratory regulation and development of vagal nerve, BCMs release from milk may prompt depression of the brain-stem respiratory centers that ultimately leads to death (Sun et al. 2003). The other investigators have claimed that the opioid peptide (BCM-7) that is rich is proline is resistant to digestion and more immature gut of infants provides enough possibilities for this peptide to be absorbed and affect various receptors in the endocrine, nervous and immune systems (Bell et al. 2006). On the other hand, the transports of these peptides out of the CNS have also been established in few animal models. Although the above studies provide various proofs in favour of the hypothesis, however more research is needed in this field. These pieces of evidence are strong but not strong enough to make final nutritional recommendations to the public in general and autism and schizophrenia patients in particular. There is an enhanced comprehension of how this opioid peptide (BCM-7) may lead not only to symptoms of autism but also to fundamental abnormal neurological development linked to the serotoninergic system (Kost et al. 2009). There are also reports available that demonstrate that children suffering from autism show digestive problems and help BCM-7 to absorb and reach circulation (Cade et al. 2000). Nevertheless, the issue is still controversial. The preliminary reports performed in animals show that BCM-7 is absorbed and crosses blood-brain barrier and causes autistic manifestation (Sun and Cade 2003). Surprisingly, the human trials involving milk elimination have showed positive inferences (Knivsberg et al. 2002), however, the same is criticized because of lack of double blind protocols.

References

Banks WA, Kastin AJ (1987) Saturable transport of peptides across the blood-brain barrier. Life Sci 41(11):1319–1338

Barbaresi WJ, Katusic SK, Voigt RG (2006) Autism: a review of the state of the science for pediatric primary health care clinicians. Arch Pediatr Adolesc Med 160(11):1167–1175

Bell SJ, Grochoski GT, Clarke AJ (2006) Health implications of milk containing beta-casein with the A2 genetic variant. Crit Rev Food Sci Nutr 46:93–100

Brantl V, Teschemacher H, Blasig J et al (1981) Opioid activities of beta-casomorphins. Life Sci 28(17):1903–1909

Brantl V, Pfeiffer A, Herz A et al (1982) Antinociceptive potencies of beta-casomorphin analogs as compared to their affinities towards mu and delta opiate receptor sites in brain and periphery. Peptides 3(5):793–797

Brooks JG (1982) Apnea of infancy and sudden infant death syndrome. Am J Dis Child 136:1012–1023

Cade R, Privette R, Fregly M et al (2000) Autism and schizophrenia: intestinal disorders. Nutr Neurosci 3:57–72

Cass H, Gringras P, March J et al (2008) Absence of exogenously derived urinary opioid peptides in children with autism. Arch Dis Child 93(9):745–750

Christison GW, Ivany K (2006) Elimination diets in autism spectrum disorders: any wheat amidst the chaff? J Dev Behav Pediatr 27(2):S162–S171

Dettmer K, Hanna D, Whetstone P et al (2007) Autism and urinary exogenous neuropeptides: development of an on-line SPE-HPLC-tandem mass spectrometry method to test the opioid excess theory. Anal Bioanal Chem 388(8):1643–1651

Dubynin VA, Malinovskaya IV, Belyaeva YA et al (2008) Delayed effect of exorphins on learning of albino rat pups. Biol Bull 35(1):43–49

Grecksch G, Schweigert C, Matthies H (1981) Evidence for analgesic activity of beta-casomorphin in rats. Neurosci Lett 27(3):325–328

Hunter LC, O'Hare A, Herron WJ et al (2003) Opioid peptides and dipeptidyl peptidase in autism. Dev Med Child Neurol 45(02):121–128

IUPHAR (2008) The IUPHAR database on receptor nomenclature and drug classification. http://www.iuphar-db.org/GPCR/ReceptorListForward

Knivsberg AM, Reichelt KL, Hoien T et al (2002) A randomized, controlled study of dietary intervention in autistic syndromes. Nutr Neurosci 5(4):251–261

Koch G, Wiedemann K, Teschemacher H (1985) Opioid activities of human β-casomorphins. Naunyn Schmiedeberg's Arch Pharmacol 331(4):351–354

Kost NV, Sokolov OY, Kurasova OB et al (2009) Beta-casomorphins-7 in infants on different type of feeding and different levels of psychomotor development. Peptides 30:1854–1860

Lindstrome LH, Nyberg F, Terenius L et al (1984) CSF and plasma beta-casomorphin-like opioid peptides in postpartum psychosis. Am J Psychiatry 141:1059–1066

Maklakova AS, Dubynin VA, Nazarenko IV et al (1995) Effects of β-casomorphin-7 on different types of learning in white rats. B Exp Biol Med 120(5):1121–1124

Maslennikova NV, Sazonova EN, Timoshin SS (2008) Effect of β-casomorphin-7 on DNA synthesis in cell populations of newborn albino rats. B Exp Biol Med 145(2):210–212

Matthies H, Stark H, Hartrodt B et al (1984) Derivatives of beta-casomorphins with high analgesic potency. Peptides 5(3):463–470

Millward C, Ferriter M, Calver S et al (2008) Gluten-and casein-free diets for autistic spectrum disorder. Cochrane Database Syst Rev 16(2):CD003498

Ohinata K, Agui S, Yoshikawa M (2007) Soymorphins, novel μ opioid peptides derived from soy β-conglycinin β-subunit, have anxiolytic activities. Biosci Biotechnol Biochem 71(10):2618–2621

Panksepp J, Normansell L, Siviy S et al (1984) Casomorphins reduce separation distress in chicks. Peptides 5(4):829–831

Paroli E (1988) Opioid peptides from food (the exorphins). World Rev Nutr Diet 55:58–97

Pasi A, Mahler H, Lansel N et al (1993) Beta-casomorphin immunoreactivity in the brain stem of the human infant. Res Commun Chem Pathol Pharmacol 80(3):305–322

Pattinson KTS (2008) Opioids and the control of respiration. Br J Anaesth 100(6):747–758

Peckys D, Landwehrmeyer GB (1999) Expression of mu, kappa, and delta opioid receptor messenger RNA in the human CNS: a 33P in situ hybridization study. Neuroscience 88(4):1093–1135

Reichelt KL, Knivsberg AM (2003) Can the pathophysiology of autism be explained by the nature of the discovered urine peptides. Nutr Neurosci 6(1):19–28

Reichelt KL, Krem J, Scott H (1990) Gluten, milk proteins and autism: dietary intervention effects on behavior and peptide secretion. J Appl Nutrition 42:1–11

Reichelt KL, Knivsberg AM, Lind G et al (1991) Probable etiology and possible treatment of childhood autism. Brain Dysfunct 4:308–319

Sakaguchi M, Koseki M, Wakamatsu M et al (2003a) Effects of β-casomorphin-5 on passive avoidance response in mice. Biosci Biotechnol Biochem 67(11):2501–2504

Sakaguchi M, Murayama K, Jinsmaa Y et al (2003b) Neurite outgrowth-stimulating activities of β-casomorphins in neuro-2a mouse neuroblastoma cells. Biosci Biotechnol Biochem 67(12):2541–2547

Sakaguchi M, Koseki M, Wakamatsu M et al (2006) Effects of systemic administration of β-casomorphin-5 on learning and memory in mice. Eur J Pharmacol 530(1–2):81–87

Sun Z, Cade R (2003) Findings in normal rats following administration of gliadorphin-7(GD-7). Peptides 24(2):321–323

Sun Z, Cade JR, Fregly MJ, Privette RM (1999) Beta-casomorphin induces Fos-like immunoreactivity in discrete brain regions relevant to schizophrenia and autism. Autism 3(1):67–823

Sun Z, Zhang Z, Wang X et al (2003) Relation of β-casomorphin to apnea in sudden infant death syndrome. Peptides 24(6):937–943

Taira T, Hilakivi LA, Aalto J et al (1990) Effect of beta-casomorphin on neonatal sleep in rats. Peptides 11(1):1–4

Wang D, Teichtahl H (2007) Opioids, steep architecture and steep-disordered breathing. Sleep Med Rev 11(1):35–46

Wittert G, Hope P, Pyle D (1996) Tissue distribution of opioid receptor gene expression in the rat. Biochem Biophys Res Commun 218(3):877–881

Zadina EJ, Hackler L, Ge LJ et al (1997) A potent and selective endogenous agonist for the mu-opiate receptor. Nature 386:499–502

Chapter 8
Critique of the Hypothesis

Abstract American nutritionists (Goldberg and coworkers) concluded that the hypothesis is misleading, confusing and more unreliable. There is no clear cut evidence for detection of this peptide in human plasma or serum on consumption of cow milk. Hartwig also couldn't detect BCM-7 in the gut of lambs in feeding trials. The EFSA Group recommended that as per the literature available, a cause-effect association between consumption of BCM-7 and aetiology of diseases like T1D, CHD and neurological disorders is not established. Therefore, an official EFSA risk valuation of these peptides is not suggested. Truswell suggested that the ecological correlations are indecisive and alone can't be applied for the formulation of dietary recommendations. The animal models (rabbits) used for the development of correlations has many drawbacks in design and experimentation.

8.1 Critics against Type 1 Diabetes Mellitus (T1D)

8.1.1 Ecological Correlations

Although there are many ecological associations that correlate consumption of A1 milk with incidence of T1D. This is further supported by few animal trials and case control studies. Nevertheless, this correlation has been subject of controversy highlighted by very few scientists. A comprehensive discussion on the criticism of these correlations is described below. The author advocating against the hypothesis suggest that the ecological correlations are indecisive and alone can't be applied for the formulation of dietary recommendations.

- The ecological data generated in some countries or regions correlate consumption of A1 milk with incidence of T1D. The quality of cow milk either A1 or A2 in these areas has been represented through breed composition. However, the critics object whether the diabetes developing subjects were really the same who consumed A1 milk.
- The authors also speculate that infants usually take cow milk in the form of infant formulas and the latter contains a little fraction of cow milk.

© Springer Nature Singapore Pte Ltd. 2020

M. R. Ul Haq, *β-Casomorphins*, https://doi.org/10.1007/978-981-15-3457-7_8

- The cow milk used for preparation of infant formulas contains more whey fraction compared to casein, the hypothesis is related to latter not the former.
- It is also quite possible that the milk used for the manufacture of infant formulas may not belong to the same country where the product is consumed.
- The genetics of the incidence of T1D was not taken into consideration. For instance, the people living in Finland who develop T1D more easily have been determined to have HLA haplotypes with increased frequency (Reijonen et al. 1991).
- While looking in at Mediterranean area, the incidence of T1D is highest in the Sardinia Island. However, the HLA types and emigration reports have demonstrated that this incidence (T1D) is determined genetically (Muntoni and Muntoni 1999).
- Jones and coworkers in 1998 have found a negative correlation between breast feeding and incidence of T1D (Jones et al. 1998). Moreover, the environmental and socioeconomic difference exists between formula-fed and breast-fed infants. Furthermore, the infants feeding on mother's milk don't have the option of taking A1 milk.
- In the field of human nutrition, health and disease, the scientists have practiced the associationships between incidence of chronic diseases and food consumption insignificant. The linkages between incidence of coronary heart disease and sugar intake (Yudkin 1964), colon cancer and meat consumption (Armstrong and Doll 1975) and breast cancer with fat intake (World Cancer Research Fund 1997) have been regarded as insignificant and spurious (FAO/WHO Expert Consultation 1998; Committee of Medical Aspects of Food and Nutrition Policy 1998; Truswell 2002).
- In the ecological studies, some of the uncertainties have been found in the form of nationwide statistics for A1 β-casein compared to total cow milk casein and even more improbabilities for these factors compared to milk intake on an average.
- In the ecological studies, no doubt some of the countries or regions were selected for the generation of correlations. However, some very significant developed, developing and emerging dairy countries were not found in the studies e.g., Ireland and Netherlands.
- The other authors pointed out towards country or region wise ecological correlation misjudgment. They highlighted that only twenty prosperous healthcare countries were chosen for building up of correlations, however, the number of such countries should be more than thirty five (Beaglehole and Jackson 2003).
- The authors however admitted that these types of correlations (ecological or epidemiological) are a starting point for building-up of hypothesis. Nevertheless, these studies should be supported by well-designed observational findings followed by more intricate and affluent clinical experiments. The animal trials in mice, rat, rabbits etc. will also help in unraveling the mechanism. However, these animal trials alone can never be sufficient for formulating consumer nutritional recommendations.

- Altman in 1991 was very true in stating that the correlations established through ecological studies at the international level are very difficult to interpret for final conclusions. These correlations must not be regarded as the fundamental ones as soon as these are backed up by collateral evidence (Altman 1991).
- The other authors in the field of 'Nutritional Epidemiology' argued that the drawbacks of these nation-wide correlations is that these are difficult to reproduce individually that is a very significant part in the biomedical research. However, the nutritional evidences may be enriched or upgraded in addition to the refinement of analysis. Nevertheless, the generated data is still not truly independent, the subjects under study, their nutritional status and the measured parameters are the same. The collection of more new ecological reports won't be too useful. Overall, ecological studies help build-up the hypothesis but not enough to offer final decisions regarding the associations between consumption of dietary elements and the incidence of diseases (Willett 1990).
- One of the significant factors for establishment of a correlation between incidence of a disease like ischaemic heart disease and nutritional parameters is the consumption of tobacco at national level. The intake of latter product through smoking is well-recognized risk factor for the aforementioned disease, however, to the very surprise; the same has been ignored by Laugesen and Elliott's in their work (Laugesen and Elliott 2003).

8.1.2   Animal Models

- The animal models mostly used in the studies are either rats (BioBreeding, BB) or mice (NOD). The genetics of these models for development of this disorder is strong like in humans (Powers 2001).
- The number of animals in an experimental group that develop diabetes differs with the usage of weaning diet. Some of the investigators observed that the soy and wheat induced diabetes in BB rats (Scott 1996). However, others demonstrated that cow milk casein or total proteins or skim milk enhanced the incidence of diabetes in the aforementioned experimental animals (Paxson et al. 1997).
- Paxson and coworkers argued that it is most probable that bovine immunoglobulin G and bovine serum albumin (BSA) might encourage damage of pancreatic islet cells without enhancing the frequency of diabetes in experimental animals (NOD mice) (Paxson et al. 1997).
- Keeping the previous discussion under consideration, the time was absolutely set to confirm or validate the results of Elliott and coworkers performed in NOD mice comprehensively (Elliott et al. 1997). With this concern a large huge three-centre animal trials were designed in London, UK and at the University of Auckland. The approach was careful and well-designed. New Zealand Dairy Research Institute formulated nine types of diets for these experimental animals and supplied to these centers for feeding of animals (Beales et al. 2000).

- At the end of feeding (250 days) in London trial, the frequency of incidence of diabetes was highest in control (without milk). In Ottawa, the animals that were fed (150 days) with control diet (mixed cereals) were induced with highest diabetes. Surprisingly, the difference in incidence of diabetes in these animals with A1 and A2 cow milk β-casein was statistical non-significant. These alternate cow milk β-casein variants were fed either with "pregestimil" or with "prosobee".

- The animal trial performed in Auckland had to be stopped half way because of the outbreak of infection (Clostridium sp.) that caused death of many animals. Till the time animals survived, the results had the pattern as observed at the other two places. Moreover, when the researchers took samples of the pancreas of these animals for histological examination, they didn't find any statistical difference in the cytokine pattern on feeding A1 or A2 cow milk β-casein.

- The esteemed peer of researchers (Beales et al. 2000) therefore concluded that the earlier claims that regard A1 cow milk β-casein variant as diabetogenic compared to A1 cow milk β-casein variant was not conformed or validated. Therefore it is a time to divert more attention towards other nutritional elements like consumption of wheat and incidence of diabetes.

- The findings in this area are confusing as to whether some additional cow milk protein may increase incidence of diabetes in these animal models. When total cow milk protein or BSA individually was supplemented to the formulated hydrolyzed diet, the results indicated non-significant changes in the incidence of diabetes in experimental animals (BB rats) (Virtanen et al. 1991; Malkani et al. 1997). Moreover, supplementation of skim milk to the formulated diet had no outcome on the incidence of diabetes in experimental animals (NOD mice) (Coleman et al. 1990; Virtanen et al. 1991).

8.1.3 *Human Case-Control Studies*

- These studies have kept under consideration the history of feeding during infancy and childhood in T1D cases. These cases were then compared with controls which had similar age, sex and social status. Quite evidently, the incidence of diabetes occurs at various ages in childhood, the particulars of feeding (infants) commonly have to be trusted on mother's recollection that is frequently 10 or 15 years before.

- In 1994, Gerstein studied diabetes incidence in eight countries and found thirteen case-control studies. He reviewed these case-control studies comprehensively and analytically. Moreover he applied meta-analysis on the studies that had the finest strategy. The results indicated that those reports that lessened the possibility for bias, the risk for incidence of T1D was around 1.5 fold higher with a history of initial exposure to cow milk or less than 3 months mothers feeding (Gerstein 1994).

- Norris and Scott performed meta-analysis on 17 studies taken from literature in addition they inquired about the consistency of initial exposure to milk. Moreover, they considered the possible bias from various reaction rates of both controls and cases. In very lesser number of cases where actual records were found instead of mother's memory data, the incidence of T1D was merely 1.13 for infants who were not breast-fed. Therefore, the authors concluded that enhanced risk for incidence of diabetes with any of the newborn nutritional exposures is minimal (Norris and Scott 1996).

- In another attempt by the same investigators, 253 children were screened from those families that have T1D. The antibody profile was assessed in terms of three serum antibodies against β-cells of islets (pancreatic). Out of all the cases studied, 18 cases were observed where infant feeding was same compared to the control ones (Norris et al. 1996).

- Other investigators selected 36 children with diabetes and the number of controls selected was seven times more than cases, the genetic vulnerability to diabetes was kept under consideration in terms of HLA screening. The similar results were observed where cases and controls that were fed with mother's milk for 2 months didn't show significant changes in incidence of diabetes (Virtanen et al. 2000).

- A small investigation carried out by Virtanen and coworkers reported that breast feeding was preventive against incidence of T1D (Virtanen et al. 1991).

- Kimpimaki and coworkers in 2001 screened infants genetically at risk for T1D in terms expression of antibodies against islet cells. The investigation was based on 65 children who were found positive for different antibodies including ICA, IAA, GADA and 1A–2A. These children were found to had early exposure to cow milk (or short duration of mothers feeding) compared to controls ($n = 390$) (Kimpimaki et al. 2001).

- It is not necessary that the expression of autoantibodies in diabetes will assure the incidence of T1D (Couzin 2003).

- A randomized controlled trial with perfect design will help solve whether breast feeding protects or cow milk exposure increases risk of incidence of T1D.

- A Trial to Reduce Insulin dependent diabetes in the Genetically at Risk (TRIGR) collected cases from USA, Europe, Canada and Australia. In this scenario, those pregnant women were chosen who were diabetic or had diabetic child or husband. The results indicated that the qualifying children took either consistent infant formula (composed of cow milk) or Nutramigen (infant formula in which large cow milk protein are cleaved) (Couzin 2003). It was also clear that the countries where these studies have been carried out demonstrate greater consumption of A1 milk. The incidence of the T1D has not been considered due to the effect of milk but rather a reduction in defense of immune system owing to the deficiency of mothers milk.

Goldberg et al. (2002), the American nutritionists concluded as follows:

"The discussion that there is an increased risk for incidence of T1D with cow milk intake in early childhood is misleading or confusing. The other claims that consider A1 β-casein as a risk factor for the incidence of the same disorder are even more unreliable. Moreover, the components present in wheat and soya seem to be more potent diabetogens compared to

cow milk (Scott 1995). Furthermore there is no clear cut evidence for detection of this peptide in human plasma or serum after taking A1/A2 β-casein variants of cow milk (Hill et al. 2002; Teschemacher 1987). The claims put forward by Elliott and coworkers in 1997 that this opioid peptide acts on lymphocytes present in the gut and through some unknown mechanism stimulates an auto-immune response to β-cells (insulin secreting cells) that ultimately lead to their damage and finally satisfied amount of hormone (insulin) isn't secreted (Elliott et al. 1997). Nevertheless, others investigators like Hartwig didn't find any release of this opioid peptide (BCM-7) in the gut of lambs in feeding trials (Hartwig et al. 1997)".

8.2 Critics against Coronary Heart Diseases (CHD)

8.2.1 Ecological Correlations

The time of experimentation taken by few investigators isn't sufficient for establishment of the correlation (McLachlan 2001; Laugesen and Elliott 2003) between mortality due to CHD and intake of A1 β-casein of cow milk. The consumption of tobacco was not taken under consideration in these correlations, since the smoking is well established risk factor for CHD, therefore these ecological studies do have lacunas. The usage of regular milk or total dairy intake between these regions and development of CHD is abandoned by all investigators as Yadkin's hypothesis of sucrose consumption and development of CHD was excluded in 1970s. One of the most recent and comprehensive work published in this area (Laugesen and Elliott 2003) involves many countries. However, all the countries don't follow the pattern of average regression line. Many countries like France and Austria consume almost similar β-casein of cow milk that is 0.93 g/day, nevertheless, the mortalities due to CHD is 33 and 88 per one lac, respectively. Many countries (Canada, Sweden, Iceland, Australia, Israel, Germany and Island of Jersey) had mortalities due to CHD that ranged between 70 and 80 per one lac, however, the intake of cow milk β-casein varied from highest in Sweden (2.8 g/day) to lowest in Jersey (0.3 g/day) (Laugesen and Elliott 2003). From last many decades, accurate regards of the mortalities due to CHD have been kept, while it has been observed that in some countries the rates have gone up (eastern Europe) while in others the rates have gone down (Australia, Finland, America). Mentioning Switzerland in this perspective will be significant enough because there is quite trustworthy statistics on consumption of milk and health status. The intake of milk protein in this country is fixed (25 g/head/day), the mortalities due to CHD have declined (from 160 in 1980 to 95 in mid-1900 per one lac per year) (Hill et al. 2003). In the last decade, Crawford and coworkers analysed the data of intake of various food components and deaths due to CHD, the data was taken from forty seven countries by WHO and FAO. The various nutritional elements under observation were cow milk protein, cheese, alcohol, coffee and meat. The correlations were developed for each individual element and deaths due to CHD for every second year from 1969 to 1995. The correlation established very wonderful, it was observed that the correlations developed with cow milk protein was 0.7 till 1983, afterwards, the same have declined to zero.

The milk protein composition for various countries regards 30–55% A1 β-casein of total β-casein. Therefore, it implies that a correlation should have been established if this variant of β-casein is really a causative agent (Crawford et al. 2003).

From the nutritional perspective, two biochemical pathways have been proposed by investigators that links CHD with intake of cow milk casein. It is assumed that consumption of aforementioned protein increases the levels of cholesterol in comparison with soya proteins in animal (rabbit) trials. However, non-significant changes have been observed in human subjects when casein milk protein is exchanged by soya protein (Van Raaij et al. 1979, 1981, 1982; Grundy and Abrams, 1983). The caseins weren't differentiated into A1 and A2 forms, but a mixture of two alternate types. Therefore it was concluded (Sacks et al. 1983) that the amount of casein that is usually consumed in diets (USA) was unable to change the lipid levels in plasma of adult subjects. Moreover, from applied clinical nutrition perspective, milk products (non-fat) can be exploited in lowering lipid levels in plasma. The concept behind this aforesaid effect was taken from a Masai tribe living in East Africa who consume large quantities of fermented milk but experience little cholesterol levels in plasma and occasionally demonstrates CHD. Certain nutritionists are of the view that free-living individuals may consume huge quantities of whole cow milk without increasing their cholesterol levels in plasma (Howard and Marks, 1977, 1979). The balanced diet of the people gets altered if they consume extremely large (several liters) quantities of milk on daily basis. With average milk intake on daily basis (1–2 L/day) in reports there is controlled-intake metabolic unit environments, it been found that the milk increases cholesterol levels in plasma and however intake of skim milk is unable to lower it (Hussi et al. 1981; Roberts et al. 1982; Howard and Marks 1982). With more advancement it is speculated that there is no apparent cholesterol decreasing element in bovine milk. As per critics, there is neither a cholesterol-lowering (aqueous phase) nor cholesterol-increasing (casein) factors that elucidate a mechanism linking A1 milk with CHD. Moreover, no possible mechanism has been demonstrated that links incidence of CHD and intake of A1/A2 milk.

A comprehensive ecological study was performed by Ness and coworker in 2002 that was comprised of 5765 adult males located in the Scotland (west) and that was followed up for 25 years. The subjected were questioned for the quantity of milk consumed per day. However, they didn't mention about the quality of milk like decreased fat milk. The results of the study showed an inverse relationship between coronary heart and stroke mortality and milk intake. The probability was that the adult males who took more milk may have better standard of living that resulted in much confusion. Therefore the findings were regarded for smoking, cholesterol levels, social status, consumption of alcohol, respiratory disease bronchitis, education, type of transportation and age. The findings revealed that adjusted relative risk (RR) in 2350 mortalities from CHD was 0.93 and 0.64 where milk intake was one and two pints of milk per day (the pattern found was statistically significant, P¼0.05) (Ness et al. 2002). In another study by the same workers, they composed seven more potential studies that correlated milk intake with cerebrovascular and coronary deaths. The authors in any of the study couldn't establish a relationship between milk consumption and deaths (Ness et al. 2002). These individual studies appear to discard the findings of McLachlan (2001) and Laugesen and Elliott (2003) that showed a correlation between A1 β-casein intake of studied countries and mortalities due to CHD.

Moreover, the WHO/FAO Consultation listed 23 nutritional elements that were associated or not connected to cardiovascular diseases. Surprisingly milk wasn't found in that list. It is quite possible that the milks in various developed countries may contain A1 β-casein (see Chap. 1), therefore it implies that this specific varaints of cow milk casein isn't viewed as a threat element for establishment of cardiovascular disease (WHO/FAO Consultation 2003).

8.2.2 Animal Studies

There is only one comprehensive animal study that demonstrates the role of A1 and A2 β-casein and development of CHD (Tailford et al. 2003). However, the opponents have criticized the study on several grounds. The challengers regard the:

- Period of trial too short i.e., merely 6 weeks.
- Numbers of animals in each experimental group non-satisfactory (6 per group).
- Rabbits used in this study not the best models for this study as these animals were given formulated diets with replaced animal proteins (20%) whose diet otherwise is leaves.
- Unrealistic approach for development of atherosclerosis in humans.
- Atherosclerosis formation occurs in many years in humans.
- Histopathological evidences from Tailford's trial were different from that develops in humans.
- Few changes were observed in cholesterol levels in plasma between experimental animal groups when fed on this formulated diet, this is however, difficult to apply to human beings.
- Difference in cholesterol levels in plasma of human being hasn't been revealed on feeding A1 or A2 cow milk β-caseins.
- Assessment of fatty streaks (aortic) wasn't performed 'blind' to the experimental group
- Difference in the zones concerning A1 and A2 cow milk β-caseins was found non-significant in aortas or carotid arteries of the experimental animals who consumed cholesterol in the composed diets (Tailford et al. 2003).

In an editorial published by Mann and Skeaff in 2003, they reveal that:

- The findings Tailford and coworkers observed suffer from remarkable limitations; therefore these results are difficult to extrapolate for humans.
- They assume that it will be irresponsible to think if they may be applied to consumer health measures.

- The more satisfactory human, epidemiological and animal trial confirmations that exist for A1 cow milk β-casein demonstrate that with increased consumption of vitamin E a reduction in cardiovascular risk was observed.
- Nevertheless, numerous huge randomized controlled experiments utilizing vitamin E additions were unable to validate substantiation of benefit (Mann and Skeaff 2003).

8.3 European Food Safety Authority Report (EFSA)

A scientific report of EFSA was published by DATEX Working Group on BCMs. They reviewed the literature comprehensively about A1 and A2 β-caseins and related peptides. They whole literature was composed into a single document comprised of more than hundred pages. They presented their views about BCMs, their production, activity and role in various human illnesses like T1D, cardiovascular diseases and neurological manifestations (EFSA 2009).

They group argues that inadequate evidence is presently available that demonstrates effect of alternate variants of cow milk β-casein with consumers health. Regarding the release of these opioid peptides from cow milk, they stated that production of this peptide from infant formulas is indecisive. However, production of few BCMs from these formulas utilizing simulated gastrointestinal digestion (SGID) has been detected. However, they further published in their report that the quantitative detection of these peptides during in vivo digestion from milk products in humans is not available. They further report that the transport of small peptides like BCMs from milk across the gut mucosa is possible. Nevertheless, the measurable data that shows these findings are lacking. The working group further admits that there are some correlations that link consumption of this variant of cow milk with some health complications like T1D, CHD and neurological manifestations. However, they are of the view that ecological correlations suffer with the problem of not establishing cause-effect associationship and these correlations are unable to accommodate for probable confusing causes. Indeed, these correlations are at their best for building up of a hypothesis. Nevertheless, they are to establish a cause-effect associationship. These correlations will become very weak when other factors will be taken into account like consumption of A1 β-casein individually, composition of β-casein variants, and the incidence of T1D in various studied countries. The amount of β-casein (A1 and B) produced in many countries with increased or decreased occurrence of T1D looks comparatively insignificant and is unable to explain the change in frequency of T1D through various countries.

The EFSA Group finally concluded and recommended that as per the literature available, a cause-effect association between consumption of BCM-7 and related peptides and aetiology of diseases like T1D, CHD and neurological disorders is not established. Therefore, an official EFSA risk valuation of these peptides is not suggested.

References

Altman DG (1991) Practical statistics for medical research. Chapman & Hall, London, p 298

Armstrong B, Doll R (1975) Environmental factors and cancer incidence in different countries with special reference to dietary practices. Int J Cancer 15:617–631

Beaglehole R, Jackson R (2003) Balancing research for new risk factors and action for the prevention of chronic diseases. NZ Med J 116:291–292

Beales PE, Elliott RB, Flohe S et al (2000) A multi-centre, blinded international trial of the effect of A1 and A2 β-casein variants on diabetes incidence in two rodent models of spontaneous Type 1 diabetes. Diabetologia 45:1240–1246

Coleman DL, Kazuva JE, Leiter EH (1990) Effect of diet on incidence of diabetes in non-obese diabetic mice. Diabetes 39:432–436

Committee of Medical Aspects of Food and Nutrition Policy (1998) Nutritional aspects of the development of cancer. In: Report of the working group on diet and cancer of the committee of medical aspects of food & nutrition policy. The Stationery Office, London

Couzin J (2003) Diabetes' brave new world. Science 300:1862–1865

Crawford RA, Boland MJ, Hill JP (2003) Changes over time in the associations between deaths due to ischaemic heart disease and some main food types. Austral J Dairy Technol 58(116):295–313

Elliott RB, Wasmuth HE, Bibby NJ et al (1997) The role of b-casein variants in the induction of insulin-dependent diabetes in the non-obese diabetic mouse and humans. In: Seminar on milk protein polymorphism. IDF Special Issue no. 9702. Int Dairy Fed, Brussels, pp 445–453

European Food Safety Authority (EFSA) (2009) Review of the potential health impact of β-Casomorphins and related peptides. Sci Rep 231:1–107

FAO/WHO Expert Consultation (1998) Carbohydrates in human nutrition. Report of a Joint FAO/WHO Expert Consultation, FAO Food and Nutrition Paper 66. Food & Agriculture Organization of the UN, Rome

Gerstein HC (1994) Cow's milk exposure and type 1 diabetes mellitus. A critical review of the clinical literature. Diabetes Care 17:13–19

Goldberg JP, Folta C, Must A (2002) Milk: can a 'good' food be so bad? Pediatrics 110:826–832

Grundy SM, Abrams J (1983) Comparison of actions of soy protein and casein on metabolism of plasma lipoproteins and humans. Am J Clin Nutr 38:245–252

Hartwig A, Teschemacher H, Lehmann W et al (1997) Influence of genetic polymorphisms in bovine milk on the occurrence of bioactive peptides. In: Milk protein polymorphism. Int Dairy Fed, Brussels, pp 459–460

Hill JP, Crawford RA, Boland MJ (2002) Milk and consumers health: a review of the evidence for a relationship between the consumption of beta casein A1 with heart disease and insulin dependent diabetes mellitus. Proc NZ Soc Animal Production 62:111–114

Hill JP, Boland M, Crawford RA et al (2003) Changes in milk consumption are not responsible for increase in the incidence of type-1 diabetes or decrease in the incidence of deaths due to ischaemic heart disease. Austral J Dairy Technol 58:188

Howard AN, Marks J (1977) Hypocholesterolaemic effect of milk. Lancet 2(8031):255–256

Howard AN, Marks J (1979) Effect of milk products on serum cholesterol. Lancet 2(8149):957

Howard AN, Marks J (1982) The lack of evidence for a hypocholesterolaemic factor in milk. Atherosclerosis 45:243–247

Hussi E, Miettinen TA, Ollus A et al (1981) Lack of serum cholesterol lowering effect of skimmed milk and butter milk under controlled conditions. Atherosclerosis 39:267–272

Jones M, Swerdlow AJ, Gill LE et al (1998) Pre-natal and early life risk factors for childhood onset diabetes mellitus: a record linkage study. J Epidemiol 27:444–449

Kimpimaki T, Erkkola M, Korhonen S et al (2001) Short-term exclusive breast-feeding predisposes young children with increased genetic risk of Type 1 diabetes to progressive beta-cell autoimmunity. Diabetologia 44:63–69

Laugesen M, Elliott R (2003) Ischaemic heart disease, Type 1 diabetes, and cow milk A1 b-casein. NZ Med J 116:295–313

Malkani S, Nomplegi D, Hanse JW et al (1997) Dietary cow's milk protein does not alter the frequency of diabetes in the BB rat. Diabetes 46:1133–1140

Mann J, Skeaff M (2003) Editorial: b-casein variants and atherosclerosis-claims are premature. Atherosclerosis 170:11–12

McLachlan CNS (2001) β-casein A1, ischaemic heart disease mortality, and other illnesses. Med Hypotheses 56:262–272

Muntoni S, Muntoni S (1999) New insights into the epidemiology of Type 1 diabetes in Mediterranean countries. Diabetes/Metab Res Rev 15:133–140

Ness AR, Smith DG, Hart C (2002) Milk, coronary heart disease and mortality. J Epidemiol Community Health 55:379–383

Norris JM, Scott FW (1996) A meta-analysis of infant diet and insulin-dependent diabetes mellitus: do biases play a role? Epidemiology 7:87–92

Norris JM, Klingensmith G, Yu L et al (1996) Lack of association between early exposure to cow's milk protein and b-cell autoimmunity. Diabetes autoimmunity study in the young (DAISY). J Am Med Assoc 276:609–614

Paxson JA, Weber JG, Kulczycki A (1997) Cow's milk-free diet does not prevent diabetes in NOD mice. Diabetes 46:1711–1717

Powers AS (2001) Diabetes mellitus. In: Braunwald E, Fauci AS, Kasper DL et al (eds) Harrison's principles of internal medicine, vol 15. McGraw Hill, New York, pp 2112–2114

Reijonen H, Ilonen J, Knip M et al (1991) HLA-DQB1 alleles and absence of Asp 57 as susceptibility factors of IDDM in Finland. Diabetes 40:1640–1644

Roberts DCK, Truswell AS, Sullivan DR et al (1982) Milk, plasma cholesterol and controls in nutritional experiments. Atherosclerosis 42:323–325

Sacks FM, Breslow JL, Wood PG et al (1983) Lack of an effect of dairy protein (casein) and soy protein on plasma cholesterol of strict vegetarians. An experiment and a critical review. J Lipid Res 24:1012–1020

Scott F (1995) AAP recommendations on cow milk, soy and early infant feeding. Pediatrics 96:515–517

Scott FW (1996) Food-induced type 1 diabetes in the BB rat. Diabetes/Metab Rev 12:341–359

Tailford KA, Berry CL, Thomas AC et al (2003) A casein variant in cow's milk is atherogenic. Atherosclerosis 170:13–19

Teschemacher H (1987) β-Casomorphins: do they have physiological significance? In: Goldman AS, Atkinson Hanson LA (eds) Human lactation 3. Springer, Boston, MA

Truswell AS (2002) Meat consumption and cancer of the large bowel. Eur J Clin Nutr 56(1): S19–S24

Van Raaij JMA, Katan MB, Hautvast JGAJ (1979) Casein, soya protein, serum cholesterol. Lancet 314(8149):958

Van Raaij JMA, Katan MB, Hautvast JGAJ et al (1981) Effects of casein versus soy protein diets on serum cholesterol and lipoproteins in young healthy volunteers. Am J Clin Nutr 34:1261–1271

Van Raaij JMA, Katan MB, West CE et al (1982) Influence of diets containing casein, soy isolate, and soy concentrate on serum cholesterol and lipoproteins in middle-aged volunteers. Am J Clin Nutr 35:925–934

Virtanen SM, Rasanen L, Aro A et al (1991) Infant feeding in children less than 7 years of age with newly diagnosed IDDM. Diabetes Care 14:415–417

Virtanen SM, Laara E, Hypponen E et al (2000) Cows milk consumption, HLA-DQBI genotype, and type 1 diabetes. A nested case-control study of siblings of children with diabetes. Diabetes 49:912–917

WHO/FAO Consultation (2003) Diet, nutrition and the prevention of chronic diseases. Report of a Joint WHO/FAO Consultation, WHO Technical Report Series 916. Word Health Organization, Geneva

Willett W (1990) Nutritional epidemiology. Oxford University Press, New York, pp 8–9

World Cancer Research Fund (1997) Total fat and breast cancer. In: Food, nutrition and the prevention of cancer: a global perspective. American Institute for Cancer Research, Washington, DC, pp 261–267

Yudkin J (1964) Diet and coronary thrombosis. Hypothesis and fact. Lancet 270(6987):155–162

Chapter 9
Conclusions and Future Perspectives

Abstract The ecological statistics, human case-control observations, in vivo and in vitro studies in addition to the biochemical and pharmacological pathways are strong enough to establish a correlation between A1 "like" milk consumption and increased risk for various health disorders. However, the criticism cited by American Nutritionists (Goldberg and coworkers), DATEX Working Group (EFSA) and Truswell can't be overlooked. The present status of the hypothesis isn't in a position to make consumer recommendations regarding consumption of cow A1 milk. More mechanistic studies involving well-designed animal and human trials are essential for validation. Moreover, cellular, molecular, biochemical and immunological mechanisms with detailed signal cascade pathways must be explored to reach at final conclusions. Overall, the hypothesis is potentially important, intriguing, fascinating, and possibly significant. However, verified and authenticated research with reproducible results is needed to make final recommendations and guidelines regarding world's dairy sector and public or consumer health.

9.1 Cow Milk

It is a colloidal solution that contains numerous biomolecules essential for nutrition, growth and maintenance and immunity of human infants and adults. Cow milk proteins are categorized into caseins and whey proteins. Caseins are phosphoproteins (precipitate at pH 4.6) and are further divided into α, β, κ-forms. The whey proteins include α- and β-lactglobulins, albumin, antibodies, lactoferrin and transferrin. Among caseins, the β-casein is composed of 209 amino acids with at least 12 genetic variants and represented as A1, A2, A3, B, C, D, E, F, H1, H2, I, G. The mutation on exon VII on sixth chromosome of cow β-casein gene leads to the substitution from cytosine to adenine that in turn leads to replacement to histidine from proline at position 67. The above-mentioned mutation is of great importance as it leads to the division of milk into A1 "like" and A2 "like". The former further includes A1, B, C, F and G alleles while latter further includes A2, A3, D, E, H1, H2 and I alleles.

© Springer Nature Singapore Pte Ltd. 2020

M. R. Ul Haq, *β-Casomorphins*, https://doi.org/10.1007/978-981-15-3457-7_9

9.2 Frequency of A1/A2 β-Casein Allele in Cows

In western countries, there is higher frequency of A1, A2 and B alleles of β-casein. Normande, Jersey, Simmental breeds show lower A1 β-casein allele frequency. However, Normande, Jersey and Hereford possess higher frequency of B allele of β-casein. Guernsey cows are reported to have almost entirely of A2 allele while Gray, Hungarian Spotted, Brown Swiss and Brown Italian cows contain both alleles. The frequency of A1 or A2 alleles in Holstein and Friesian cows is region-dependent. Indian indigenous cattle possess predominantly A2 allele, however, the crossbred cattle contain both A1 and A2 alleles. Sahiwal possess higher A2 allele frequency compared to A1. Tharparkar and buffalo (Murrah) showed predominance of A2 allele while the river buffalo indicated only A2 allele. The crossbred cows (Karan Fries and Karan Swiss) possesses both alleles but still with higher frequency of A2 allele. Ongole (Zebu cattle) and crossbred Frieswal possess remarkably very low frequency of A1 compared to A2 allele. The frequency of A1 allele is very low in Malnad Gidda, Kasargod variety and Jersey. Rathi and Kankrej cattle have almost entirely of A2 allele. Badri cattle were found to have increased A2 allele frequency compared to A1 allele of β-casein of cow milk.

9.3 A1/A2 Cow Milk Hypothesis

A1/A2 cow milk hypothesis was established by Elliot, McLachlan and coworkers during early 90s. They suggested that A1 "like" cow milk has a considerable role to play in increasing the risk for the incidence of various health complications including type T1D, CHD, schizophrenia and autism. It is established that a seven (7) amino acid peptide is released from only A1 "like" varaints and demonstrate "morphine" like activity. This peptide is actually correlated with increased risk for the incidence of these disorders. Many ecological studies, few animal trials, some case control reports and very few in vitro evaluations support this hypothesis. Biochemistry of BCM-7 release, pharmacology (opioid nature of these peptides), subjective experiences, milk allergies and milk intolerances further back this hypothesis. On the other hand, DATEX Working Group (EFSA report) and Truswell's arguments advocate against this hypothesis. The opponents regard this as inadequate data and mere suggestive evidence and refute these well-established cause-effect correlations (EFSA 2009; Truswell 2005). The evidences in favour and critical analysis of the hypothesis and further my at least one decade's expertise in the research area put me in a state to conclude that:

The generated ecological data, human case-control findings, in vivo and in vitro studies followed by biochemical and pharmacological mechanisms are strong enough to establish a correlation between A1 "like" milk consumption and increased risk for various health disorders.

However, the counter arguments furnished by American Nutritionists, DATEX Working Group (EFSA) and Truswell can't be ignored. The present status of the hypothesis isn't in a position to make consumer or public recommendations regarding consumption of cow A1 milk.

Therefore, I suggest that more mechanistic studies involving well-designed animal and human trials are essential for validation. Moreover, cellular, molecular, biochemical and immunological mechanisms with detailed signal cascade pathways must be explored to reach at final conclusions. Overall, the hypothesis is potentially important, intriguing, fascinating, and possibly significant. However, verified and authenticated research with reproducible results is needed to make final recommendations and guidelines regarding world's dairy sector and public or consumer health.

9.4   Structure, Classification and Production of BCMs

BCMs are the opioid peptides that bind μ-receptors and vary in number and sequence of amino acids. BCM-7 is the most abundant with prolonged effect while BCM-5 is the most active opioid peptide. The common characteristic feature among BCMs is the presence of tyrosine at the N-terminal and another aromatic amino acid residue (Phe or Tyr) at third or fourth position (Kostyra et al. 2004). Cieslinska and coworkers in 2007 reported the detection of BCM-7 in whole or raw cow milk (Cieslinska et al. 2007). Cathepsin B has been found to release BCM-10 from the β-casein of cow milk (Considine et al. 2004). It was Hamel and co-researchers for the first time in 1985 who detected BCM immunoreactive substances from milk incubated with lactic acid bacteria (LAB) (Hamel et al. 1985). The wild type *Lactobacillus helveticus* L89 either wild form or mutant stains (pepX deficient) were found to breakdown BCM-7 to BCM-4 fully or partially at 37 °C incubated for 120 min. The more advanced techniques like LC/MS further demonstrated the production of BCM-4 but not BCM-7 in pasteurized milk (65 °C for 30 min) (Matar and Goulet 1996). Schieber and Brückner in 2000 reported the presence of BCM containing peptide sequences from 57–68 and 57–72 from A1 variants of β-casein with *L. delbruekii* ssp. *bulgaricus* Lb1466 and *S. thermophilus* St1342. When fermented milk was digested with pepsin and trypsin, peptides similar to BCM-11 and BCM-4 were detected (Rokka et al. 1997). Jarmolowska and coworkers in 1999 reported the production of BCM-7 from Brie cheese samples and the content obtained was from 5 to 15 mg/kg cheese (Jarmolowska et al. 1999). Earlier, Muehlenkamp and Warthesen in 1996 couldn't find any release of BCMs from the cheese samples used by the aforementioned studies. BCM-9 has been detected in Cheddar cheese, the latter secreted by LAB in the same cheese is found to have little or no pepX activity (Singh et al. 1997). There are other reports available that couldn't detect presence of BCM-7 in these cheeses but the long peptide precursor sequences incorporated in it. Accordingly, Addeo and coworkers in 1992 demonstrated the presence of peptide precursors (20–21 amino acid residues) possessing in it BCM-7 sequence in ripened Parmigianino Reggiano cheese. The presence of BCM-3 but not BCM-5 was detected in Edam and probiotic Edam cheeses manufactured from Bifidobacterium with HPLC/UV detection system (Sabikhi and Mathur 2001). Jarmolowska in 2007 detected four peptides in the infant formula "Humana" in the extract of pepsin and pepsin-trypsin hydroxylate. The peptides detected were BCM-5, casoxin C and 6 and lactoferroxin A (Jarmolowska et al. 2007). In 2008, the production of BCM-5 and BCM-7 was analyzed during SGID in commercial and experimental milk-based infant formulas (De Noni 2008). The significance of pepsin in SGID of β-casein was well demonstrated by Jinsmaa and Yoshikawa in 1999 where production of BCM-9, −13 and −21 or (Val59)-BCMs occurred with pepsin hydrolysis of the Leu58-Val59 linkage in β-casein. Further production of BCM-7 (17 mmol/mol β-casein) in little amounts was observed with elastase-leucine aminopeptidase hydrolysis without the preliminary pepsin hydrolysis. Moreover, there wasn't any production of BCM-7 when either chymotrypsin or trypsin was employed instead of pancreatin and further pepsin (individually) was also unable to release BCM-7. In 1999, Macaud and co-researchers detected release of BCM-3 (no BCM-7) on pepsin hydrolysis (Macaud et al. 1999). In 2008, De Noni and co-workers found that the B variant of β-casein depicted highest BCM-7 release (5–176 mmol/mol casein), this was followed by release from A1 genetic variant. The release of BCM-7 from A2 genetic variant couldn't be elucidated. Moreover, the release of BCM-5 couldn't be detected from either of the genetic variants of β-casein or change in pH during SGID (De Noni 2008). The release of BCM-7 from both the A1 and A2 variants of β-casein was observed by Cieslinska et al. (2007) from unprocessed milk, however, the release from the former was almost four times higher compared to latter after digesting these with pepsin at pH 2.0 for 24 h (Cieslinska et al. 2007). The BCM-7 was recovered, however, the release of this peptide was not quantified in the infant formula (reconstituted and UF permeate) that was digested with pepsin (pH, 3.5) followed by hydrolysis with Corolase PPTM (Hernández-Ledesma et al. 2004). However, the same researcher couldn't detect the same peptide in pepsin digested (pH, 3.5) infant formula and pancreatin (porcine) (Hernández-Ledesma et al. 2007). Nevertheless, the peptide BCM-9 was found in the same infant formula by UV-HPLC. Jarmolowska and coworkers in 2007 observed the release of BCM-5 in the infant formula digested by pepsin and pepsin-trypsin. The yield of BCM-5 in IF extracts (0.39 μg/mL) and its hydrolysate (43.11 μg/mL) was also observed. However, there wasn't any production of BCM-7 in the extracts before or after the SGID (Jarmolowska et al. 2007). We also checked in our laboratory the release of BCM-7 and BCM-5 from A1A1, A1A2 and A2A2 genetic variants of β-casein. The amount of BCM-7 released from A1A1 was found to be 0.20 ± 0.02 mg/g β-casein that was almost 3.2 times more than heterozygous A1A2 variant of β-casein. On the other hand, as expected, the release of BCM-7 wasn't observed from A2A2 β-casein variant. Moreover, the release of BCM-5 wasn't detected from any of the variants of β-casein (Raies et al. 2015).

9.5   Physiological Activity of BCMs

BCMs have a role in the transport of amino acids (Brandsch et al. 1994) and secretion of mucus in the intestine (Trompette et al. 2003; Zoghbi et al. 2006). Further they affect the postprandial metabolism (Takahashi et al. 1997), reduction in the suppression of intake of fats (White et al. 2000), increase in the gastrointestinal transit time (Mihatsch et al. 2005) and modulation in the absorption of electrolytes and water (Daniel et al. 1990). A remarkable inhibition in the gastric transit, up-regulation of DDPIV and myeloperoxidase activity in the intestine was observed on consumption of A1 β-casein (Barnett et al. 2014). Moreover, intake of A1 milk was related to elevated inflammatory response of gastrointestinal system, deterioration of PD3 symptoms, the delay in gastro-intestinal transit, and the reduced cognitive processing speed and accuracy (Jianqin et al. 2016). Bovine BCM-5 and BCM-7 demonstrated a depression in respiratory frequency and tidal volume (Hedner and Hedner 1987). BCMs (BCM-4, 5, 6 and morphiceptin) showed sedative activity in addition to analgesic effects (Paroli 1988; Matthies et al. 1984; Zadina et al. 1987). BCMs showed potent cardiovascular effects (Holaday 1983), produced hypotension and bradycardia (Widy-Tyszkiewicz et al. 1986). BCM-5 produced a bipolar effect, ionotropic activity at lower and cardiodepressive response at higher doses (Liebmann et al. 1986). BCM-7 has been observed to secrete mucin from the jejunum of rats (Trompette et al. 2003) through expression of mucins (Zoghbi et al. 2006). Suppression in lymphocyte proliferation (taken from lamina propria) was observed with opioid peptides at lower concentration (as low as 10 mM) (Elitsur and Luk 1991). BCM-7 and BCM-10 showed a proliferation of lymphocytes at lower and stimulation at higher doses (Meisel and Bockelmann 1999). BCM-7 was found to secrete histamine from peripheral leukocytes (humans) and resulted in the wheal and flare response in the skin and the allergic responses (Kurek et al. 1992). BCM-7 caused histamine release from peritoneal cells in vitro and induced wheal formation and obstruction in the bronchia in sensitized animals (guinea pigs) (Kurek and Malaczynska 1999).

9.6   A1 Milk and Health Concerns

During the last few decades the consumption of A1 milk has been linked with increased risk for various health complications like T1D, CVD and neurological manifestations (described in previous chapters). It is assumed that BCM-7 released from A1 milk is actually the hypothetical risk factor. The ecological, human case-control reports, few animal and in vitro trials support this hypothesis. The biochemical studies elucidating the release of BCM-7, opioid and binding properties of this peptide resembling that of morphine revealed through pharmacological investigations support this hypothesis. The immunological findings observed through enhanced antibody profile in non-obese diabetic mice and bio-breed rats against A1 milk further establishes a role of this type of protein in the incidence of these complications. BCM-7 with opioid properties released from A1 milk decreases the lymphocyte proliferation derived from human intestine in vitro. Therefore, there is enough probability that such an immunosuppression may affect the progression of gut-linked immune tolerance and in turn subdues the defense against the enteroviruses (Laugesen and Elliott 2003a, b; Elliott et al. 1997). In our laboratory, we found an inflammatory response of A1 β-casein and commercially synthesized BCM-7 and BCM-5 in mice models. We observed an increase in the expression of inflammatory markers like myeloperoxidase, monocyte chemotactic protein, interleukin-4 and production of histamine. Moreover an enhanced antibody profile was observed through production of IgE, IgG, IgG1/IgG2a and associated expression of TLR-2 and TLR-4 and leucocyte infiltration in mice intestine. These observations demonstrate that intake of A1 β-casein and BCM-7 and BCM-5 induce inflammatory immune response in mice intestine most likely through Th2 pathway (Raies et al. 2014a, b).

9.7 Critics of A1/A2 Hypothesis

- *American Nutritionists (Goldberg et al. 2002)*

- *"The discussion that there is an increased risk for incidence of T1D with cow milk intake in early childhood is misleading or confusing. The other claims that consider A1 β-casein as a risk factor for the incidence of the same disorder are even more unreliable. Moreover, the components present in wheat and soya seem to be more potent diabetogens compared to cow milk (Scott 1995). Furthermore there is no clear cut evidence for detection of this peptide in human plasma or serum after taking A1/A2 β-casein variants of cow milk (Hill et al. 2002; Teschemacher et al. 1986). The claims put forward by Elliott and coworkers in 1997 that this opioid peptide acts on lymphocytes present in the gut and through some unknown mechanism stimulates an auto-immune response to β-cells (insulin secreting cells) that ultimately lead to their damage and finally satisfied amount of hormone (insulin) isn't secreted (Elliott et al. 1997). Nevertheless, others investigators like Hartwig didn't find any release of this opioid peptide (BCM-7) in the gut of lambs in feeding trials (Hartwig et al. 1997).*
- *EFSA Report 2009*
- *The EFSA Group finally concluded and recommended that as per the literature available, a cause-effect association between consumption of BCM-7 and related peptides and aetiology of diseases like T1D, CHD and neurological disorders is not established. Therefore, an official EFSA risk evaluation of these peptides is not suggested.*
- *Truswell's Arguments*

- *The correlation between consumption of A1 milk and increased risk for incidence of various complications through ecological correlations is merely suggestive evidence. There isn't satisfactory substantiation that this milk is hypothetical risk factor. The ecological correlations aren't too strong in examining causes of health complications, for instance, they didn't show a positive associationship of smoking (tobacco intake) with CHD (Laugesen and Elliott 2003a, b). The suggestions relating neurological manifestations with A1 or A2 β-caseins in cow milk is more theoretical and the suggestion is even more speculative compared to T1D and CHD.*

9.8 Overall Conclusion

The ecological data, human case-controls, in vivo and in vitro trials correlate consumption of A1 milk with increased risk for incidence of various health disorders (T1D, CHD and neurological manifestations). Moreover, biochemical and pharmacological studies in addition to the immunological perspective further support the hypothesis.

However, the criticism furnished by DATEX Working Group (EFSA), Truswell commentary and American Nutritionists can't be ignored. Therefore, the present status of the hypothesis isn't in a position to make consumer or public recommendations regarding consumption of cow A1 milk.

9.9 Future Recommendations

It is suggested that more mechanistic studies involving well-designed animal and human trials are essential for validation. Moreover, cellular, molecular, biochemical and immunological mechanisms with detailed signal cascade pathways must be explored to reach at final conclusions.

Overall, the hypothesis is potentially important, intriguing, fascinating, and possibly significant. However, verified and authenticated research with reproducible results is needed to make final recommendations and guidelines regarding world's dairy sector and public or consumer health.

If/when the hypothesis is proven correct, public or consumers may demand to lessen or eliminate this variant from their regular diet. The world dairy has to make adjustments in the milk products by lessening this variant of β-casein in the products. Alternately, the favoured variant can be increased through selective culling of A1 milk cows and by selective retention of A2 calves.

References

Barnett MP, McNabb WC, Roy NC et al (2014) Dietary a1 β-casein affects gastrointestinal transit time, dipeptidyl peptidase-4 activity, and inflammatory status relative to A2 β-casein in Wistar rats. Int J Food Sci Nutr 65:720–727

Brandsch M, Brust P, Neubert K et al (1994) β-Casomorphins-chemical signals of intestinal transport systems. In: Brantl V, Teschemacher H (eds) β-Casomorphins and related peptides: recent developments. VCH, Weinheim, pp 207–219

Cieslinska A, Kaminski S, Kostyra E et al (2007) β-Casomorphin-7 in raw and hydrolyzed milk derived from cows of alternative β-casein genotypes. Milchwissenschaft 62:125–127

Considine T, Healy Á, Kelly AL et al (2004) Hydrolysis of bovine caseins by cathepsin b, a cysteine proteinase indigenous to milk. Int Dairy J 14(2):117–124

Daniel H, Vohwinkel M, Rehner G (1990) Effect of casein and β-casomorphins on gastrointestinal motility in rats. J Nutr 120(3):252–257

De Noni I (2008) Release of beta-casomorphins 5 and 7 during simulated gastro-intestinal digestion of bovine beta-casein variants and milk-based infant formulas. Food Chem 110(4):897–903

Elitsur Y, Luk GD (1991) Beta-casomorphin (BCM) and human colonic lamina propria lymphocyte proliferation. Clin Exp Immunol 85:493–497

Elliott RB, Wasmuth HE, Bibby NJ et al (1997) The role of b-casein variants in the induction of insulin-dependent diabetes in the non-obese diabetic mouse and humans. In: Seminar on milk protein polymorphism. IDF Special Issue no. 9702. Int Dairy Fed, Brussels, pp 445–453

European Food Safety Authority (EFSA) (2009) Review of the potential health impact of β-casomorphins and related peptides. Sci Rep 231:1–107

Goldberg JP, Folta C, Must A (2002) Milk: can a 'good' food be so bad? Pediatrics 110:826–832

Hamel U, Kielwein G, Teschemacher H (1985) Beta-casomorphin immunoreactive materials in cow's milk incubated with various bacterial species. J Dairy Res 52(1):139–148

Hartwig A, Teschemacher H, Lehmann W et al (1997) Influence of genetic polymorphisms in bovine milk on the occurrence of bioactive peptides. In: Milk protein polymorphism. Int Dairy Fed, Brussels, pp 459–460

Hedner J, Hedner T (1987) Beta-casomorphins induce apnea and irregular breathing in adult rats and newborn rabbits. Life Sci 41(20):2303–2312

Hernández-Ledesma B, Amigo L, Ramos M et al (2004) Release of angiotensin converting enzyme-inhibitory peptides by simulated gastrointestinal digestion of infant formulas. Int Dairy J 14(10):889–898

Hernández-Ledesma B, Quirós A, Amigo L et al (2007) Identification of bioactive peptides after digestion of human milk and infant formula with pepsin and pancreatin. Int Dairy J 17(1):42–49

Hill JP, Crawford RA, Boland MJ (2002) Milk and consumers health: a review of the evidence for a relationship between the consumption of beta casein A1 with heart disease and insulin dependent diabetes mellitus. Proc NZ Soc Anim Product 62:111–114

Holaday JW (1983) Cardiovascular effects of endogenous opiate systems. Annu Rev Pharmacol Toxicol 23:541–549

Jarmolowska B, Kostyra E, Krawczuk S et al (1999) β-Casomorphin-7 isolated from Brie cheese. J Sci Food Agric 79:1788–1792

Jarmolowska B, Szlapka-Sienkiewicz E, Kostyra E et al (2007) Opioid activity of humana formula for newborns. J Sci Food Agric 87(12):2247–2250

Jianqin S, Leiming X, Lu X et al (2016) Effects of milk containing only A2 beta casein versus milk containing both A1 and A2beta casein proteins on gastrointestinal physiology, symptoms of discomfort, and cognitive behavior of people with self-reported intolerance to traditional cows' milk. Nutr J 15(1):45

Jinsmaa Y, Yoshikawa M (1999) Enzymatic release of neocasomorphin and β-casomorphin from bovine β-casein. Peptides 20(8):957–962

Kostyra E, Sienkiewicz-Szapka E, Jarmolowska B et al (2004) Opioid peptides derived from milk proteins. Pol J Food Nutr 13(54):25–35

Kurek M, Malaczynska T (1999) Food allergy, atopic dermatitis-nutritive casein formula elicits pseudoallergic skin reactions by prick testing. Int Arch Allergy Immunol 118(2):228–229

Kurek M, Przybilla B, Hermann K et al (1992) A naturally occurring opioid peptide from cow's milk, beta-casomorphine-7, is a direct histamine releaser in man. Int Arch Allergy Immunol 97:115–120

Laugesen M, Elliott R (2003a) Ischaemic heart disease, type 1 diabetes, and cow milk A1 β-casein. N Z Med J 116(1168):U295

Laugesen M, Elliott R (2003b) The influence of consumption of A1 β-casein on heart disease and type 1 diabetes–the authors reply. Z Med J 116(1170):U367

Liebmann C, Barth A, Neubert K et al (1986) Effects of β-Casomorphin on 3h-ouabain binding to guinea-pig heart membranes. Pharamzie 41:670–671

Macaud C, Zhao Q, Ricart G et al (1999) Rapid detection of a casomorphin and new casomorphin-like peptide from a peptic casein hydrolysate by spectral comparison and second order derivative spectroscopy during HPLC analysis. J Liq Chromatogr Relat Technol 22(3):401–418

Matar C, Goulet J (1996) β-Casomorphin-4 from milk fermented by a mutant of lactobacillus helveticus. Int Dairy J 6(4):383–397

Matthies H, Stark H, Hartrodt B (1984) Derivatives of β-casomorphins with high analgesic potency. Peptides 5:463–470

Meisel H, Bockelmann W (1999) Bioactive peptides encrypted in milk proteins: proteolytic activation and thropho-functional properties. Antonie Van Leeuwenhoek 76(1):207–215

Mihatsch WA, Franz AR, Kuhnt B et al (2005) Hydrolysis of casein accelerates gastrointestinal transit via reduction of opioid receptor agonists released from casein in rats. Biol Neonate 87(3):160–163

Muehlenkamp MR, Warthesen JJ (1996) Beta-casomorphins: analysis in cheese and susceptibility to proteolytic enzymes from Lactococcus lactis ssp Cremoris. J Dairy Sci 79(1):20–26

Paroli E (1988) Opioid peptides from food (exorphins). World Rev Nutr Diet 55:58–63

Raies MH, Kapila R, Sharma R et al (2014a) Comparative evaluation of cow β-casein variants (A1/A2) consumption on th2-mediated inflammatory response in mouse gut. Eur J Nutr 53(4):1039–1049

Raies MH, Kapila R, Saliganti V (2014b) Consumption of β-casomorphins-7/5 induce inflammatory immune response in mice gut through th2 pathway. J Funct Foods 8:150–160

Raies MH, Kapila R, Kapila S (2015) Release of β-casomorphin-7/5 during simulated gastrointestinal digestion of milk β-casein variants from Indian crossbred cattle (Karan Fries). Food Chem 1(168):70–79

Rokka EL, Syväoja J, Tuominen KH (1997) Release of bioactive peptides by enzymatic proteolysis of lactobacillus GG fermented UHT milk. Milchwissenschaft 52:675–678

Sabikhi L, Mathur BN (2001) Qualitative and quantitative analysis of β-casomorphins in Edam Cheese. Milchwissenschaft 56(4):198–200

Schieber A, Brückner H (2000) Characterization of oligo- and polypeptides isolated from yoghurt. Eur Food Res Technol 5:310–313

Scott F (1995) AAP recommendations on cow milk, soy and early infant feeding. Pediatrics 96:515–517

Singh TK, Fox PF, Healy A (1997) Isolation and identification of further peptides in the diafiltration retentate of the water-soluble fraction of cheddar cheese. J Dairy Res 64(3):433–443

Takahashi M, Moriguchi S, Suganuma H et al (1997) Identification of casoxin C, an ileum-contracting peptide derived from bovine kappa-casein, as an agonist for C3a receptors. Peptides 18(3):329–336

Teschemacher H, Umbach M, Hamel U et al (1986) No evidence for the presence of β-casomorphins in human plasma after ingestion of cow's milk or milk products. J Dairy Res 53:135–138

Trompette A, Claustre J, Caillon F et al (2003) Milk bioactive peptides and β-casomorphins induce mucus release in rat jejunum. J Nutr 133(11):3499–3503

Truswell AS (2005) The A2 milk case: a critical review. Eur J Clin Nutr 59:623–631
White CL, Bray GA, York DA (2000) Intragastric beta-casomorphin (1-7) attenuates the suppression of fat intake by enterostatin. Peptides 21(9):1377–1381
Widy-Tyszkiewicz E, Czonkowski A, Szreniawski Z (1986) cardiovascular response to morphiceptin in spontaneously hypertensive and normotensive rats. Pol J Pharmacol Pharm 38:51–56
Zadina JE, Kastin AJ, Manasco PK et al (1987) Long-term hyper analgesia induced by neonatal β-endorphin and morphiceptin is blocked by neonatal Tyr-MIF-1. Brain Res 409:10–18
Zoghbi S, Trompette A, Claustre J et al (2006) β-Casomorphin-7 regulates the secretion and expression of gastrointestinal mucins through a μ-opioid pathway. Am J Physiol Gastrointest Liver Physiol 290(6):1105–1113